T0232616

FIELD GUIDE TO ECOSITES
OF THE MID-BOREAL ECOREGIONS
OF SASKATCHEWAN

*J.D. Beckingham[1], D.G. Nielsen[2],
and V.A. Futoransky[3]*

SPECIAL REPORT 6

Canadian Forest Service
Northwest Region
Northern Forestry Centre
1996

This project was funded by
Natural Resources Canada, Canadian Forest Service under the
Canada–Saskatchewan Partnership Agreement in Forestry

[1,2,3] Geographic Dynamics Corp., 10368B - 60 Avenue, Edmonton, Alberta, T6H 1G9

© Her Majesty the Queen in Right of Canada 1996
Catalogue No. Fo29-34/6-1996E
ISBN 0-660-16387-X
ISSN 1188-7419

This publication may be purchased from:

UBC Press
University of British Columbia
6344 Memorial Road
Vancouver British Columbia V6T 1Z2
    Phone (604) 822-5959
    Fax 1-800-668-0821

A microfiche edition of this publication may be purchased from:

Micromedia Ltd.
240 Catherine Street, Suite 305
Ottawa, Ontario K2P 2G8

Canadian Cataloguing in Publication Data

Beckingham, John D. (John David), 1961-

Field guide to ecosites of the mid-boreal ecoregions of Saskatchewan

(Special report, ISSN 1188-7419 ; 5)
Includes an abstract in French.
Includes bibliographical references.
ISBN 0-660-16387-X
Cat. no. Fo29-34/6-1996E

1. Forest ecology — Saskatchewan — Classification — Handbooks,
manuals, etc.
2. Forest site quality — Saskatchewan — Handbooks, manuals, etc.
I. Nielsen, D.G.; Futoransky, V.A.
II. Northern Forestry Centre (Canada)
III. Series: Special report (Northern Forestry Centre (Canada)) ; 6.

QH541.F6B42 1996   574.5'2642'097123   C96-980007-X

Beckingham, J.D.; Nielsen, D.G.; Futoransky, V.A. 1996. *Field guide to ecosites of the mid-boreal ecoregions of Saskatchewan*. Nat. Resour. Can., Can. For. Serv., Northwest Reg., North. For. Cent., Edmonton, Alberta. Spec. Rep. 6.

ABSTRACT

An ecological classification system was developed for the mid-boreal ecoregions of Saskatchewan through the analysis of vegetation, soil, site, and forest productivity information. The hierarchical classification system has three levels—ecosite, ecosite phase, and plant community type. Thirteen ecosites are described with further detail provided by subdivision into ecosite phase and plant community type. A soil type classification system that describes 17 soil types was also developed. Management interpretations were made for drought, excess moisture, soil rutting hazard, compaction hazard, puddling hazard, soil erosion hazard, frost heave hazard, soil temperature limitations, vegetation competition, windthrow hazard, productivity, and season of harvest. Color photos and drawings for 103 common plants of the mid-boreal ecoregions of Saskatchewan are presented.

RÉSUMÉ

Un système de classification écologique a été mis au point pour les écorégions mi-boréales de la Saskatchewan à partir des résultats de l'analyse de la végétation, des sols, des sites et des informations sur la productivité forestière. Le système de classification hiérarchique comporte trois niveaux : l'aire écologique, la phase de l'aire et le type de communauté végétale. Treize aires écologiques sont décrites et leur subdivision en fonction de la phase et du type de communauté végétale permet de donner de plus amples détails. Un système de classification pédologique décrivant 17 types de sols a également été élaboré. Des interprétations ont été faites pour la gestion des sécheresses, des excès d'eau, des dangers de formation d'ornières, de compactage du sol, de formation de flaques d'eau, d'érosion du sol et de soulèvement par le gel, des limites dues à la température du sol, de la compétition végétale, des dangers de déracinement par le vent, de la productivité et de la saison de récolte. Des photos couleurs et des dessins de 103 plantes communes des écorégions mi-boréales de la Saskatchewan sont présentés.

ACKNOWLEDGMENTS

This field guide was made possible with funding from the Canada–Saskatchewan Partnership Agreement in Forestry. The authors thank Program Managers Steve Price and Vic Begrand for their support throughout this project. Dr. Ian Corns, Scientific Advisor, and Philip Loseth, Project Facilitator, both of the Canadian Forest Service, are acknowledged for their expertise and contributions.

The direction and knowledge of the Site Classification Steering Committee contributed significantly to the development of this field guide and is greatly appreciated. Committee members include, from Saskatchewan Environment and Resource Management (SERM), David Lindenas (Forestry Branch), Robert Wright (Parks and Facilities Branch), and Terry Rock (Wildlife Branch); from Prince Albert Model Forest, Thomas Bouman; from Weyerhaeuser Canada Ltd. Saskatchewan, Paul LeBlanc; from Mistik Management Ltd., Mike Martel and Ken Broughton; and from Saskfor MacMillan Limited Partnership, John Daisley.

The authors are indebted to the staff of Geographic Dynamics Corp. who were dedicated to this project from start to finish. Their contributions were crucial to every component of the project from experimental design, data collection, and analysis through to electronic publishing. Many people helped including Linda Kershaw, David Chesterman, David Mussell, Margo Jarvis, Rick Reid, Brian Sinclair, Tony Szumigalski, Corey De La Mare, Marge Meijer, Barb Tumm, Andrew Wilson, Carla Lehman, Jack Morgan, Tara Howarth, Michael Chajkowski, and Kirsten Giles.

Dr. Don Pluth of the University of Alberta, Richard Sims of the Canadian Forest Service, Sault Ste. Marie, and Glen Padbury of the Centre for Land and Biological Resources Research, Research Branch, Agriculture and Agri-Food Canada, Saskatoon, reviewed the final drafts of the field guide. This review was important in the evolution of the document to its final form and is appreciated. Suggestions from Drs. Wayne Strong,

Ellen Macdonald, Peter Achuff, Dave McNabb, Vic Lieffers, and Howard Anderson, all of whom reviewed similar documents for Alberta, are acknowledged because many of their suggestions were applicable to and improved the Saskatchewan field guide.

The contribution of data from Alberta Land and Forest Service is appreciated. Harry Archibald and Grant Klappstein of Alberta Land and Forest Service and David Presslee of Weldwood of Canada Ltd. worked in association with the authors on similar field guides in Alberta where many concepts and ideas developed. This information has strengthened the Saskatchewan classification system and their contribution is appreciated.

Several people assisted in the verification of plant species identification. The assistance of Dr. Dale Vitt, René Belland, and Bernard Goffinet of the University of Alberta and Derek Johnson of the Canadian Forest Service is acknowledged.

Information was provided for the field guide by Genny Greif of SERM, Wildlife Branch, Sheila Lamont from Saskatchewan Conservation Data Centre, and Burke Korol of University of Saskatchewan. Their contributions are acknowledged.

Special thanks are extended to Dennis Lee of the Canadian Forest Service and to Gordon Barth for their excellent plant drawings. We also thank the Ontario Ministry of Natural Resources for drawings by Annalee McColm and Jane Bowles, the British Columbia Provincial Museum for a drawing by Patricia Drukker-Brammall, and the University of Washington Press for a drawing by John Rumley. The technical assistance of Doug Allan, Canadian Forest Service, is appreciated.

The following friends and colleagues provided photographs for this publication: Peter Achuff, Doug Allan, Lorna Allen, Rick Annas, Frank Boas, Ian Corns, Ray Coupé, Helen Habgood, Ron Hall, Ted Hogg,

Julie Hrapko, Derek Johnson, Dave Mussel, Robert Norton, Glen Padbury, Jim Pojar, Don Thomas, and Rob Wright. Their generosity is acknowledged.

Brenda Laishley and Elaine Schiewe, Northern Forestry Centre Publications, are acknowledged for their contribution; their attention to detail greatly improved the final product.

This publication was prepared and printed with funds provided by the Canada–Saskatchewan Partnership Agreement in Forestry.

CONTENTS

APPENDIXES

INDEX OF COMMON AND SCIENTIFIC NAMES OF ILLUSTRATED PLANTS

FIGURES

TABLES

1.0 INTRODUCTION

With increased demands placed upon the forested lands of Saskatchewan, appropriate management has become increasingly important. The most appropriate management involves balancing the utilization of various resources at a sustainable level without degradation of alternate resource values—preservation, wildlife, recreation, mining, logging, grazing, etc. Integrated and sustainable resource management requires a thorough understanding of ecosystem dynamics. Ecosystems are complex and evolving systems with the flow of energy and matter determined by the interaction of climate, landforms, topography, soils, vegetation, animals, and all other organisms.

Ecosystem classification assists with the organization of our current understanding about ecosystem function. This organization is achieved by grouping research plots into similar and functional units that respond to disturbance in a similar and predictable manner.

The ecosystem classification system outlined in this document organizes ecological information into a format that helps understanding and provides a structure for ecologically based management. The system provides a framework for organizing, overlaying, and updating forest land management interpretations. For example, it can be applied as part of an operational cruise, pre-harvest assessment procedure, or for silvicultural planning and/or wildlife management planning. The system has been developed primarily as a field tool that should be used in conjunction with the knowledge of the user. The objectives of the ecosite classification system are:

- to facilitate the application of ecological information to decisions on a wide variety of activities included within the realm of forest resource management;
- to facilitate the collection and organization of information to expedite the development of resource management applications and decision support systems;
- to promote communication among resource managers and between managers and the public;
- to provide a common basis for integrated planning, and;
- to reduce resource management costs by integrating ecological information into the decision-making process.

This guide contains 15 major sections, eight appendixes, and an index. Following this introduction is a brief description (Section 2.0) about the ecology of the Mid-Boreal Ecoregions of Saskatchewan. Section 3.0 provides a general explanation about the methods used to develop the classification system. Section 4.0 explains the structure of the ecological classification system, and Section 5.0 describes how to apply the system. A breakdown on how to read the fact sheets is provided in Section 6.0. Keys to the ecological units of the hierarchical system are provided in Section 7.0, followed by fact sheets for each unit in Section 8.0. Section 9.0 presents a soil type classification system with keys and fact sheets. Section 10.0 contains photographs of soil profiles representing several common soil types. Section 11.0 consists of plant photos and drawings to assist in species identification. A brief discussion about management interpretations including supporting information is provided in Section 12.0. Section 13.0 presents forest mensuration information and the relationship of the field guide to the Saskatchewan provincial forest inventory. The literature used for development of the field guide is listed in Section 14.0 followed by a glossary of terms (Section 15.0). The appendixes provide information to assist in applying the classification system. An index of common and scientific names of illustrated plants is found at the back of the field guide.

2.0 STUDY AREA

This field guide covers the Mid-Boreal ecoregions of Saskatchewan which can be divided into the Mid-Boreal Upland (MBU) and Mid-Boreal Lowland (MBL) ecoregions (Figure 1). The MBU and MBL ecoregions have been subdivided into landscape areas based on differences in physiography, surface expression, and the proportion and distribution of soils and plant community types within the area (Figure 2) (Padbury and Acton 1994). The MBU includes landscape areas in central and western Saskatchewan (E1 to E24) immediately south of the Canadian Shield. It also includes several outlying landscape areas known as the Bronson Upland (E25) and the Thickwood Upland (E26) in the west, and the Pasquia Escarpment (E27), the Pasquia Plateau (E28), the Porcupine Hills (E29), and Duck Mountain (E30) landscape areas in the east. The MBL Ecoregion consists of a low-lying level plain in east-central Saskatchewan.

2.1 Physiography

The Mid-Boreal ecoregions of Saskatchewan occur within the Interior Plains Physiographic Region of North America (Ellis and Clayton 1970). These plains comprise a large area of relatively low relief extending from the mountainous Cordillera in the west, east, and north to the Precambrian Shield to the north and east.

The Mid-Boreal ecoregions of Saskatchewan were last glaciated during the Wisconsinan glaciation (11 400 B.P.) when the Laurentide ice sheet advanced from the northeast. This glacial episode left in its wake a vast plain that contains flat-topped high hill regions, rolling upland areas, both undulating and flat lowland surfaces, and lowland regions studded with shallow lakes. The MBU Ecoregion has considerable physiographic variability while the MBL Ecoregion consists of a flat plain with very little relief. Numerous hilly regions of the MBU Ecoregion that extend above the surrounding plains were present long before the ice age and are the remnants of older plains. The abrasive and erosive power of the Laurentide ice sheet rounded and smoothed these elevated regions.

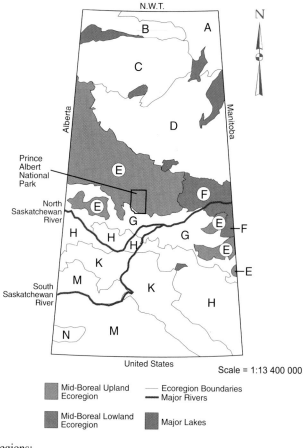

Figure 1. **Study area location.** This guide covers the Mid-Boreal Upland (E) and Mid-Boreal Lowland (F) ecoregions.

Scale = 1:13 400 000

Legend:
- Mid-Boreal Upland Ecoregion
- Mid-Boreal Lowland Ecoregion
- Ecoregion Boundaries
- Major Rivers
- Major Lakes

Ecoregions:

A = Selwyn Lake Upland
B = Tazin Lake Upland
C = Athabasca Plain
D = Churchill River Upland
E = Mid-Boreal Upland

F = Mid-Boreal Lowland
G = Boreal Transition
H = Aspen Parkland
K = Moist Mixed Grassland
M = Mixed Grassland
N = Cypress Upland

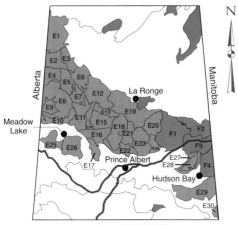

Scale = 1:12 500 000

Mid-Boreal Upland:

E1 = Firebag Hills (0)
E2 = Garson Lake Plain (0)
E3 = Palmbere Plain (2)
E4 = Christina Plain (0)
E5 = Dillon Plain (30)
E6 = Ile-à-la-Crosse Plain (5)
E7 = Canoe Lake Lowland (2)
E8 = Mostoos Upland (0)
E9 = Primrose Plain (0)
E10 = Mostoos Escarpment (72)
E11 = Waterhen Plain (59)
E12 = La Plonge Plain (30)
E13 = Mahigan Lake Plain (0)
E14 = Dore Lake Lowland (9)
E15 = Smoothstone Plain (14)

E16 = Clarke Lake Plain (41)
E17 = Leoville Hills (5)
E18 = Waskesiu Upland (29)
E19 = La Ronge Lowland (19)
E20 = Wapawekka Upland (16)
E21 = Montreal Lake Plain (27)
E22 = Emma Lake Upland (27)
E23 = Whiteswan Upland (39)
E24 = White Gull Plain (3)
E25 = Bronson Upland (45)
E26 = Thickwood Upland (17)
E27 = Pasquia Escarpment (6)
E28 = Pasquia Plateau (0)
E29 = Porcupine Hills (6)
E30 = Duck Mountain (1)

Mid-Boreal Lowland:

F1 = Mossy River Plain (0)
F2 = Namew Lake Upland (0)

F3 = Saskatchewan Delta (9)
F4 = Overflowing River Lowland (0)

Figure 2. **Landscape areas of the Mid-Boreal ecoregions of Saskatchewan.** The number of plots sampled in each landscape area is shown in parentheses. Plot distributions not shown: ecologically equivalent upland areas in Alberta (167); Mid-Boreal Upland (42) and Lowland (28) areas in Manitoba.

2.2 Climate

The climate of the MBU and MBL ecoregions of Saskatchewan can be described as cool continental, sub-arid to sub-humid, with long cold winters and warm summers with marked differences between day and night temperatures. Weather variations are influenced by three large air masses that can dominate the Mid-Boreal ecoregions at various times during the year. During the winter, the continental polar air mass brings in cold, dry air, while cool, moist air from the Pacific can be influential at any time during the year. Occasionally, the weather of Saskatchewan's Mid-Boreal ecoregions is influenced by the warm, moist air masses that originate over the Atlantic or Gulf of Mexico.

Temperatures in the Mid-Boreal ecoregions range from about -40°C to 32°C with a frost free period of generally around 80 to 100 days in the south to 70 days in the north. The mean annual daily temperature is near -0.2°C. July and August are generally the warmest months of the year during which time mean maximum temperatures exceed 20°C three out of four days, and exceed 25°C three to four days out of 10 (Bauer 1976). The MBU appears to have cooler summers and warmer winters than the MBL Ecoregion (Table 1).

Table 1. **Climate data for the Mid-Boreal ecoregions**
(Atmospheric Environment Service 1993)

Ecoregion	Total annual precipitation (mm)	Annual snowfall (cm)	Mean July temperature (°C)	Mean January temperature (°C)	Water deficit (mm)
Mid-Boreal Upland*	456	147	16.3	-18.9	180
Mid-Boreal Lowland**	452	170	17.7	-21.4	169

* Waskesiu Lake station.
** The Pas station.

Average annual precipitation ranges from 400 to 500 mm, with upland areas receiving greater than average amounts. Approximately 70% of the annual precipitation occurs as rain, the majority of which falls during June, July, and August (Atmospheric Environment Service 1993). The MBU and MBL have similar amounts of precipitation, although the MBU appears to receive a greater proportion as rain (Table 1) and the MBL has a greater proportion as snow.

The Mid-Boreal ecoregions have numerous mesoclimates due to the variable landscape of rolling uplands interspersed with basins, subdued lowlands, and deeply incised river valleys.

2.3   Soils

The soils in central Saskatchewan, as in other regions, result from the interaction of climate, parent materials, topography, organisms (primarily vegetation), and time. The soil Great Groups and Subgroups briefly described herein are based on *The Canadian system of soil classification* (Agriculture Canada Expert Committee on Soil Survey 1987) (Appendix 3). The dominant soils of the Mid-Boreal ecoregions of Saskatchewan are Organic, Gray Luvisols, and Brunisols.

Orthic Gray, Brunisolic Gray, and Gleyed Gray Luvisols are all common. Eutric Brunisols are more prevalent than the more acidic Dystric Brunisols. Orthic Gray Luvisols are prominent in the southern parkland transition area while Gleysolic and Organic soils occur where drainage is inhibited. Regosols are found where conditions are harsh or disturbance is severe (e.g., along rivers). Cryosols are soils influenced by permanently frozen horizons and occur in some bogs and poor fens where the organic matter insulates the soil from solar radiation.

2.4   Vegetation

The vegetation of the boreal forest is highly dynamic and strongly influenced by fire. There are many different interpretations of seral stages and developmental sequences in the boreal forest but all stress the dominant influence of fire and ecological adaptations to it. While the composition and abundance of plant species on a site is important to our understanding the dynamics of the ecological system, trees are the most conspicuous of the vegetational layers so their ecology will be discussed briefly.

There are 11 common tree species in the Mid-Boreal ecoregions of Saskatchewan. Eight are found in both the MBU and MBL ecoregions– aspen (*Populus tremuloides*), balsam poplar (*Populus balsamifera*), jack pine (*Pinus banksiana*), black spruce (*Picea mariana*), white spruce (*Picea glauca*), white birch (*Betula papyrifera*), tamarack (*Larix laricina*), and balsam fir (*Abies balsamea*). An additional three species occur exclusively in the MBL Ecoregion—Manitoba maple (*Acer negundo*), green ash (*Fraxinus pennsylvanica*), and white elm (*Ulmus americana*). Dix and Swan (1971) and Carleton and Maycock (1981) both suggest that tree species occurrences, associations, and co-occurrences are the result of edaphic factors and substratum moisture regimes.

Pure jack pine stands are frequently associated with dry sandy soils, aspen with loamy soils and intermediate moisture status, and black spruce with moist soils (Lee 1924; Rowe 1956; Martin 1959; Mueller-Dombois 1964; Dix and Swan 1971). Balsam poplar and black spruce are chiefly located on poorly drained sites while aspen and white spruce are favored by intermediate moisture conditions (Dix and Swan 1971). Carleton and Maycock (1981) state that certain canopy types such as tamarack and balsam poplar occur on sites with noticeable imports of nutrients in the form of lateral subsurface flow or seasonal flooding and sediment deposition. This is also true of the green ash, white elm, and Manitoba maple, which are commonly found on the floodplains of the MBL Ecoregion. These sites support richer floras than jack pine and black spruce-dominated stands, which are associated with tight ecosystem nutrient cycles (Carleton and Maycock 1981).

Swan and Dix (1966), measuring vegetation and soils along a continuum in the boreal mixedwood forest of Saskatchewan (MBU), concluded that tree species occurrences could be used to predict with reasonable accuracy the important herb species found in a stand. They found approximately one half of the understory taxa to be specific to a particular canopy type. Carleton and Maycock (1981) also studied understory canopy affinities in the boreal mixedwood forest, south of James Bay, Quebec/Ontario and reported 121 of 410 understory taxa showed specificity to the canopy classes identified.

In the Mid-Boreal ecoregions of Saskatchewan, the tree canopy is an important indicator of site conditions, influences understory species composition and abundance, and affects the availability and cycling of nutrients on a site. Forest stand classification is an important component in ecological evaluation.

Several authors have developed regional forest classification systems for Saskatchewan. Halliday (1937) delineated forest regions based on forest communities comparable to the climatic "climaxes" of Clements (1928). Rowe (1959, 1972) refined Halliday's regions and subdivided them into forest sections based on the consistent presence of particular forest associations. Saskatchewan Environment and Resource Management, Forestry Branch has a progressively evolving ecologically based forest stand classification and mapping system that realizes the indicator value of tree species and facilitates application of ecological principles on an operational level (e.g., Frey 1981; Lindenas 1985).

While wetland sites may or may not have a tree canopy, the presence, absence, and composition of plant species on a site provide important information about environmental and ecological conditions.

2.5   Wildlife

In the Mid-Boreal ecoregions of Saskatchewan, wildlife are abundant and diverse with moose, woodland caribou, mule deer, white-tailed deer, elk, black bear, timber wolf, and beaver being among the most prominent mammals. White-throated sparrow, American redstart, ovenbird, hermit thrush, and bufflehead are typical birds (Padbury and Acton 1994).

3.0   METHODS

3.1   Approach

This field guide was the product of several years of development. With the formation of the Saskatchewan "Site Classification Steering Committee" in 1992, it was deemed important that all topographic positions be integral components of the site classification system, including lowlands (bog, poor fen, rich fen, marsh, and gully). This was largely a result of the diversity of interests and objectives represented by the multidisciplinary steering committee. The committee managed the project in part through the solicitation of input from stakeholders through workshops and public forums.

Data were collected specifically for the site classification project during the 1993, 1994, and 1995 field seasons by Geographic Dynamics Corp. (GDC) (402 plots). GDC obtained plot data from Mistik Management Ltd. (111 plots) and from the Alberta provincial government for adjacent, ecologically equivalent areas in Alberta (167 plots). Data were also obtained from the Duck Mountain ecosystem classification pilot project in Manitoba (70 plots) (Knapik et al. 1988). A portion of the forest mensuration data was provided by the Forestry Branch of Saskatchewan Environment and Resource Management. A total of 750 plots were used to develop the classification for the Mid-Boreal ecoregions of Saskatchewan.

The plots that were described for this classification project were selected using a methodology consistent with the sampling concept "subjective without preconceived bias" advocated by Mueller-Dombois and Ellenburg (1974). Thus, plots were chosen to represent the typical site conditions, canopy cover, and understory species distribution for the selected stand.

Existing forest site classification systems were reviewed and many ideas were used to increase the functionality of this ecological classification system. Kabzems et al. (1986) provided considerable background into the ecology of Saskatchewan's boreal forest. The classification structure of the current guide also incorporated ideas and approaches from the *Field guide to forest ecosystems of west-central Alberta* (Corns and Annas 1986), several biogeoclimatic field guides from British Columbia (e.g., DeLong et al. 1990), and the *Field guide to the forest ecosystem classification for northwestern Ontario* (Sims et al. 1989).

This classification system evolved in close association with the *Field guide to ecosites of northern Alberta* (Beckingham and Archibald 1996) and other Alberta field guides. The parallel development of these systems in both provinces has improved the utility of the classifications as they were scrutinized, reviewed, and tested by numerous stakeholders.

Detailed vegetation, soil, site, and mensuration data were collected from each plot. The plots were selected to cover the variability in the plant communities, soils, forest cover, landforms, and topography of the Mid-Boreal ecoregions of Saskatchewan.

3.2　Classification methods

Ecological units were defined through the analysis of vegetation, site, soil, and tree production data. The data were statistically analyzed subsequent to field data collection in 1993, 1994, and 1995. Several draft reports were produced and results were presented at public workshops. Draft copies of the field guide were tested during the 1994 and 1995 field seasons by industry, governments, and consultants. The public workshops provided a structure for review of the system and a forum for the development of management interpretations related to the ecological units. These mechanisms of evaluation provided considerable insight and feedback to identify data gaps, strengthen the classification system, evaluate the keys, and outline management interpretations within the ecological classification framework. A general outline of the methods is presented.

Data from approximately 750 plots were statistically analyzed. Data were stratified into subsets based on the dominant tree species cover type. The following groups were used:

- · pine dominated;
- · deciduous tree dominated;
- · white spruce and fir dominated;
- · black spruce and tamarack dominated, and;
- · non-forested.

The classification process then involved a search for floristic similarities among plots using the SAS-PC CLUSTER procedure (SAS Institute Inc. 1990). Several cluster methods were initially tested to determine the best "raw" computer classification. The cluster methods that were tested included complete linkage, average linkage, and Wards minimum variance methods. Each clustering method was used with squared distance data (SAS Institute Inc. 1990). Average linkage appeared to produce the best classification so it was used for further data analysis. Further analysis involved weighting indicator species and incorporating moisture regime and nutrient regime values as continuous variables in the cluster analysis.

Computer classification groups that resulted from cluster analysis of each of the five dominant tree cover groups were used as the basis for further classification analysis. Vegetational and environmental data were summarized for each provisional group. The summarized data were evaluated to determine whether there were significant differences in environmental characteristics among the provisional types, and whether there were any data trends not evident from the previous analyses. The General Linear Models procedure (GLM) available with SAS was used to perform ANOVAs, regressions, means, and least squared means difference tests, and Levene nonparametric tests to determine significant differences in moisture regime, soil drainage, nutrient regime, and site index among the provisional ecological groups (Sabin and Stafford 1990; SAS Institute Inc. 1990). This process provided ecological insight to facilitate the classification process. With these analyses, ecological understanding of the Mid-Boreal ecoregions of Saskatchewan increased and the present structure of the classification system evolved.

Principal component, canonical correlation, canonical correspondence, and correspondence analysis ordination techniques were used to provide a mathematical evaluation and visual impression of ecological differences between the ecological units (Ter Braak 1988; SAS Institute Inc. 1990). As well, ecological units were plotted on a moisture/nutrient grid (edatope) to provide a visual representation of their degree of similarity. The classification groups were modified so plots classified as the same ecological unit were consistent with respect to environmental characteristics and site index.

In addition to the hierarchical integrated classification system, a soil type classification system was developed. Soil types are taxonomic units used to stratify soils based on soil moisture regime, effective soil texture, organic matter thickness, and solum depth. The system was developed to cover the variability in the soils represented by the sample plot data.

Soil types from the *Field guide to the forest ecosystems of northwestern Ontario* (Sims et al. 1989) were reviewed for their applicability in Saskatchewan. With modification, moisture and texture classes were found to be useful. The moisture levels very dry, dry, moist, and wet were considered to simplify moisture regime without losing resolution or functionality. Organic soils were separated from wet soil types because they represented functionally different units. Organic soils are classified and defined using *The Canadian system of soil classification* (Agriculture Canada Expert Committee on Soil Survey 1987). Six distinct groups (very dry, dry, moist, wet, organic, and shallow) were adopted for this field guide. These groups were further refined by attaching soil textural classes or organic matter thickness codes to the moisture regime classes. Natural breaks or divisions in moisture, texture, and organic matter depth were determined through data analysis, a review of the literature, and expert opinion.

Several soil attributes were defined as soil type modifiers to provide more resolution in the soil type classification system. These were selected based on analysis of variance and regression models and by expert opinion. Humus form and depth of organic matter were used as soil type modifiers because they had statistically significant influences ($P < 0.05$) on site index values for at least one tree species. Surface coarse fragment content and surface texture were defined as soil type modifiers because they were deemed important soil attributes due to their influence on site productivity and forest management.

4.0   SYSTEM OVERVIEW

Saskatchewan's ecosystem classification system organizes ecological information by classifying, identifying, and naming distinct ecological units to increase our understanding of ecological function and to provide a structure for ecologically based management. Saskatchewan's ecosystem classification system consists of an integrated hierarchical ecological classification with three levels (ecosite, ecosite phase, and plant community type) and a separate soil classification (Figure 3). The integrated hierarchical ecological classification and soil classification systems are nested within Saskatchewan's geographically based ecoregion classification system (Padbury and Acton 1994). A description of each level in the classification is provided below and the code and name of each ecological unit in the classification system is displayed on pages 8-3 through 8-7.

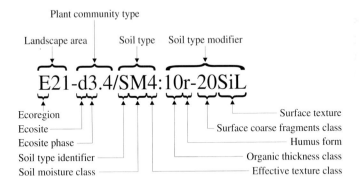

Figure 3.   Mapping codes for the hierarchical ecological classification system.

Ecoregion (e.g., E Mid-Boreal Upland)

In Saskatchewan, an ecoregion can be thought of as "an area characterized by a distinctive regional climate as expressed by vegetation" (Lacate 1969). Reference ecosites are sites where the ecosystem is more strongly influenced by the regional climate than by edaphic (soil) or landscape factors. Reference sites are usually defined as having deep, well to moderately well drained soils of medium texture that neither lack nor have an abundance of soil nutrients or moisture, and are neither exposed to nor protected from climatic extremes (Strong and Leggat 1991). The term "modal" (Strong and Leggat 1991) and "zonal" (Pojar et al. 1991) have been used as alternative terms to describe reference sites. The current field guide was designed for use in the Mid-Boreal Upland and Mid-Boreal Lowland ecoregions of Saskatchewan (Padbury and Acton 1994) (Figure 1). A capital letter is used to denote the ecoregion.

Landscape Area (e.g., E21 Montreal Lake Plain)

A landscape area is a subdivision of the ecoregion based on differences in physiography, surface expression, and the proportion and distribution of soils and plant community types within an area. A landscape area has a recurring pattern of slope, landform, soils, vegetation, and climate. A capital letter and a number are used to denote each landscape area (Padbury and Acton 1994) (Figure 2).

Ecosite (e.g., d low-bush cranberry)

Ecosites are ecological units that develop under similar environmental influences (climate, moisture regime, and nutrient regime). They are groups of one or more ecosite phases that occur within the same portion of the edatope (moisture/nutrient grid) (page 8-2). Each of the 13 ecosites are designated with a small letter, with "a" representing the driest ecosite and "m" the wettest (page 8-2). Each ecosite has been given a name that attempts to convey some information about the ecology of the unit. Ecosites are frequently named after plant species that are common or typical of the ecosite (e.g., d low-bush cranberry). The plant that the ecosite is named after, however, may not be present in every plot or stand belonging to that ecosite. This convention of ecosite naming was adopted from the site series level of British Columbia's biogeoclimatic classification system (Pojar et al. 1991). Ideally, the name for an ecosite would incorporate features of physiography and soils, but in reality that is impractical (Pojar et al. 1991) due to the high variability of ecological systems.

Ecosite, in this classification system, is a functional unit defined by moisture and nutrient regime. It is not tied to specific landforms or plant communities as in other systems (Lacate 1969), but is based on the combined interaction of biophysical factors which together dictate the availability of moisture and nutrients for plant growth. Thus, ecosites differ in their moisture regime and/or nutrient regime.

Ecosite Phase (e.g., d3 low-bush cranberry tA-wS)

An ecosite phase is a subdivision of the ecosite based on the dominant species in the canopy. On lowland sites where a tree canopy may or may not be present, the tallest structural vegetation layer with a percent cover greater than 5 determines the ecosite phase. For example, the bog ecosite (j) has a treed and a shrubby ecosite phase.

Generally, ecosite phases are believed to be mappable units. They are identified by the ecosite letter (e.g., d) and name (e.g., low-bush cranberry) along with a number (e.g., 3) representing the phase within the ecosite (Figure 3).

Differences in ecosite phases of the same ecosite, although determined largely by the dominant species in the canopy, may be expressed as differences in lower strata plant species abundance and pedogenic processes. Ecosite phases, however, have a distinct range in canopy composition and lower strata floristics. The composition of the tree canopy as well as indicating environmental conditions (Lee 1924; Rowe 1956; Martin 1959; Mueller-Dombois 1964; Dix and Swan 1971; Carleton and Maycock 1981), influences structure, diversity, composition, and abundance of understory vegetation (Moss 1953a; Rowe 1956; Dix and Swan 1971; Ellis 1986). The tree canopy and canopy-dependent factors such as degree of shading and understory species composition interact to dictate the type and quantity of organic matter, its rate of decomposition, and a site's nutrient availability. The ecosite phase level of the classification system, while being defined by canopy composition or structure, has a strong ecological basis. For forested sites the ecosite phase correlates well with forest cover types on provincial forest inventory maintenance maps (Section 13.0).

Plant Community Type (e.g., d3.4 tA-wS/low-bush cranberry–prickly rose)

Ecosite phases may be subdivided into plant community types, which are the lowest taxonomic unit in the classification system. The environmental characteristics of the community are defined at the ecosite level and to a lesser degree at the ecosite phase level. While plant community types of the same ecosite phase share vegetational similarities, they differ in their understory species composition and abundance. As one proceeds down the taxonomic levels of the hierarchy, from ecosite to ecosite phase to plant community type, the ecological variability within a unit is expected to decrease.

Plant community types are considered difficult to map from air photographs but may be important to wildlife, recreation, or other resource sectors. Each plant community type is identified by the ecosite letter, the phase number followed by a period, and another number that together constitute the plant community type (e.g., d3.4 tA-wS/low-bush cranberry–prickly rose) (Figure 3). Generally, plant community types are named by combining the name of the dominant plant species in several structural vegetation layers (e.g., a1.2 jP/blueberry/lichen).

Soil Type

Soil types are taxonomic units used to classify soils based on soil moisture regime, effective soil texture, organic matter thickness, and solum depth. Soil types can be used independently, in association with the hierarchical classification system (ecosite, ecosite phase, and plant community type), and to classify disturbed sites. The soil type is represented by a two- or three-character code. When used with the hierarchical system, the soil type is separated from it with a slash (/) (Figure 3). The first letter in the code is an S (soil type identifier) followed by a capital V, D, M, W, R, or S which represent very dry, dry, moist, wet, organic, and shallow soils, respectively. The third character of the soil type code, when required, is a number that represents the effective texture class in the very dry, dry, and moist soil types, or a letter that represents peaty (p) with the moist and wet soil types or mineral (m) with the wet soil types. The soil type classification is explained in Section 9.0.

Soil Type Modifier

Soil type modifiers are used in association with the soil type as "open legend" modifiers to provide more resolution. Soil attributes were used as soil type modifiers if they significantly influenced tree productivity or were considered to have important implications for management. Soil type modifiers include organic matter thickness, humus form, surface coarse fragment content, and surface texture (see page 9-6). The codes and classes for soil type modifiers are described in Section 9.0.

The user is encouraged to choose the level in the hierarchy required for his purposes. All levels, however, must be consulted to correctly classify a site.

5.0 APPLICATION OF THE CLASSIFICATION SYSTEM

This section describes the application of the ecosystem classification system. Ecological site identification consists of comparing field characteristics with information presented in this guide to correctly identify a site. This guide includes several tools to help identify ecosites, ecosite phases, plant community types, and soil types. There are dichotomous keys, flowcharts, edatopic grids, landscape profiles, and fact sheets to help the user apply the classification system and identify units.

As experience is gained in identifying ecological units and an understanding of ecosystem function and the interrelationships among ecosystem components is achieved, reliance on these tools diminishes. Each site, however, must be thoroughly examined so the present state of the ecosystem can be explained in terms of the interaction of ecosystem components. Combining ecological site classification with the collection of site, soil, and vegetation data provides the most complete ecological description of the site.

The most widely used survey data form for recording ecological data at a reconnaissance level in Alberta can easily be used for Saskatchewan. This form is included in the *Pre-harvest ecological assessment handbook* (Strong 1994). Forms for detailed data collection can be found in the *Ecological land survey site description manual* (Alberta Environmental Protection 1994). Various versions of these forms were used to collect data for the development of this field guide.

Plant species abundance, organic matter thickness, and soil moisture vary across the landscape as one ecological unit grades into another. There is considerable natural variability in the landscape with ecotones being common. Thus, not every site encountered will be easily classified into the ecological units described in this field guide. This field guide is intended to promote ecological thinking and a better understanding and appreciation of the complexity and interrelatedness of ecosystems.

The steps that describe how to use the field guide are outlined (Figure 4) and details provided for each step, so the system can be applied correctly. The user needs to have some basic knowledge about ecosystem classification concepts, soil description, and plant identification. The following 10 steps provide a framework for the application of the ecological classification system.

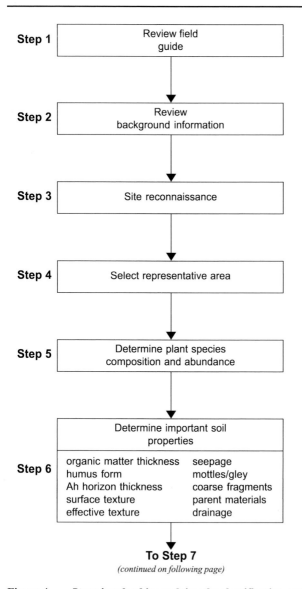

Step 1 Review field guide

Step 2 Review background information

Step 3 Site reconnaissance

Step 4 Select representative area

Step 5 Determine plant species composition and abundance

Step 6 Determine important soil properties

organic matter thickness	seepage
humus form	mottles/gley
Ah horizon thickness	coarse fragments
surface texture	parent materials
effective texture	drainage

To Step 7
(continued on following page)

Figure 4. Steps involved in applying the classification system.

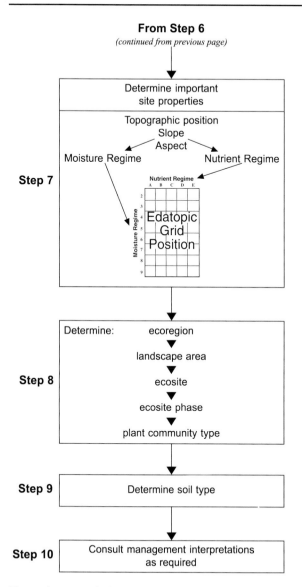

Figure 4. concluded.

5.1 STEP 1:
Review this field guide to become familiar with the classification system and its structure

It is important to become familiar with the classification system so it can be applied correctly. Part of this process requires working through the 10 steps described here, but other sections are helpful for increasing your understanding of the classification system. Please read System Overview (Section 4.0), How to Read the Fact Sheets (Section 6.0), and other sections including the use of the keys (Section 7.0).

5.2 STEP 2:
Review background information and pre-stratify the area to be classified

Information about the area you are investigating should be reviewed to learn what you can about the landscape and ecology. Soil surveys (including maps) (Appendix 6) and forest inventory maintenance maps are available for most of the area encompassed by the Mid-Boreal ecoregions of Saskatchewan. Use existing maps, reports, and aerial photographs to pre-stratify the area into ecological units of relatively uniform landform, topographic position, and forest cover type.

5.3 STEP 3:
Carry out a quick reconnaissance of the site to be classified

While walking around the site take note of the topography, position on the landscape, and general plant species distribution including trees and understory. Other things to look for include evidence of past disturbance, effects of disease or insect infestations, and the homogeneity and size of the site to be classified.

5.4 STEP 4: Choose a location that appears to be representative of the area to be classified

Locate an area for your assessment that appears to be representative of the site to be classified, and is as homogeneous in slope, plant cover, and overstory canopy conditions as possible. Avoid locating the sample plot in areas that have received significant natural or artificial disturbance unless those are the areas you are planning to sample. Also avoid ecotone areas or relatively small areas that are transitional between homogeneous ecological units such as slope breaks, landform boundaries, forest cover type boundaries, and road edges, unless the intent is to sample those areas to satisfy specific studies or management objectives.

5.5 STEP 5: Determine the plant species composition and abundance

Determine the plant species composition and abundance within a 10×10 m plot. Also record any species that appear to be representative of the ecological unit but occur outside the plot. Make sure to indicate which species occurred outside the plot. Abundance is estimated by determining the amount of ground area that is covered by the plant species when its canopy is projected onto the ground surface. Within a 10×10 m plot, 1 m² is equivalent to 1% cover. An illustrative guide is provided to assist with estimates of proportions and may help in estimating percent cover (Figure 5).

There are numerous guides to assist in plant species identification. In addition to the plant pictures in Section 11.0, several books and guides that the authors have found useful are listed in Appendix 8. Nomenclature in this field guide follows Farrar (1995) for trees, Moss (1983) for other vascular plants, Ireland et al. (1987) and Schofield (1992) for mosses, Stotler and Crandall-Stotler (1977) for liverworts, and Egan (1987) for lichens.

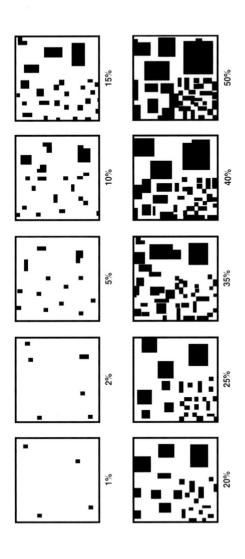

Figure 5. **Guide for estimating percent vegetation cover and proportions of mottles or coarse fragments.** Each quarter of any one square has the same amount of black (*after* Corns and Annas 1986).

5.6   STEP 6:
Determine important soil properties

To collect soils data, a soil pit must be dug or augered. In most cases a soil pit 60 cm deep will be adequate but there are exceptions. A deeper pit is required when the soil has a coarse to moderately coarse texture. In these cases, the pit is dug deeper to see if there are finer-textured layers that are influencing ecological function below the 60 cm of coarse to moderately coarse material. A deeper pit is also required when the plant community type on the site cannot be explained by the site conditions and soil conditions above 60 cm. This type of assessment, however, takes experience and will come with time. If detailed soils information is required for research, then the pit should be dug to a depth of 1 m or until parent materials or bedrock are reached (C horizon). The minimum soils information that should be collected at a site to classify it correctly is outlined below.

Organic matter thickness: Organic matter thickness is the depth of surface organic horizons in centimetres.

Humus form: Humus forms consist of soil horizons located at or near the surface of a pedon, which have formed from organic residues either separate from, or intermixed with mineral materials (Klinka et al. 1981). Thus, a humus form can comprise entirely organic, or both organic and mineral horizons (see Section 5.6.1).

Ah horizon thickness: The thickness of the Ah horizon (defined in Appendix 2 and the glossary) is measured in centimetres.

Surface texture: The dominant soil texture within the top 20 cm of mineral soil is the surface texture (see Section 5.6.2).

Effective texture: Effective texture for mineral soils is generally defined as the textural class of the finest-textured horizon that occurs 20 to 60 cm below the mineral soil surface and is at least 10 cm thick. The effective texture for organic soils is the dominant organic matter decomposition class (fibric, mesic, or humic) 40 to 80 cm below the surface (Appendix 1).

Seepage: Seepage is the movement of water to a site caused by gravitational forces through subsurface flow from areas higher in the landscape. Evidence of seepage could include the occurrence of water seeps, wet ground, or hydrophytic plants. Hydrophytic plants can occur on sites that appear to be better drained.

Mottles: Mottles appear as spots or blotches that contrast with the dominant soil color or matrix color. These are areas of intense oxidation of iron in once-saturated (reducing conditions) mineral soils. Mottles appear orange to red in color (see Section 5.6.3).

Gley: Gley can be seen as a distinct gray to blue color in mineral soil horizons and is indicative of long periods of water saturation. Due to the anaerobic conditions, minerals in these horizons have been reduced and the soils may emit a rotten egg (hydrogen sulfide) odor (see Section 5.6.3).

Coarse fragments: Coarse fragments are expressed as the percentage of the volume of the fragments larger than 2 mm in diameter within the soil. Coarse fragments affect forestry operations, site development, and plant production since high percentages of coarse fragments reduce the volume of the rooting medium.

Parent material/landform: The parent material is the surficial material from which soils are formed. Classification of parent material follows that outlined in *The Canadian system of soil classification* (Agriculture Canada Expert Committee on Soil Survey 1987).

Drainage: Soil drainage refers to the rate of water removal from the soil relative to supply. Soil drainage reflects the potential occurrence of soil water in excess of field capacity. The prolonged occurrence of free water can result in anaerobic conditions that can create mottles and gleying in the soil (see Section 5.6.5)

5.6.1   Humus form classification

A humus form is a group of soil horizons located at or near the surface of a pedon, which have formed from organic residues, either separate from, or intermixed with mineral materials (Klinka et al. 1981). Thus, a humus form can comprise entirely organic, or both organic and mineral horizons. The mineral horizons that are included in humus forms are restricted to melanized A horizons characterized by a significant accumulation of organic matter from residues of root systems or the activity of soil fauna, both of which may be associated with infiltration (Klinka et al. 1981). Humus horizons, in relation to the rest of the soil profile, are characterized by intensive biological activity.

The von Post scale of decomposition is used to estimate the degree of decomposition in organic soil materials (peat) in wetlands (Appendix 1). Soil humus forms are described below, while a dichotomous key to the classification of humus forms is presented in Figure 6.

Mor: The mor humus form has diagnostic F and H horizons, with a distinct boundary evident between the organic and mineral layer. There is little or no intermixing of organic and mineral horizons.

Moder: Diagnostic organic horizons of the moder humus form have varying degrees of intermixing between the organic and mineral horizons, producing a gradual transition between the horizons.

Mull: In the mull humus form the diagnostic F and H horizons are commonly lacking. There is considerable mixing of organic material into the surface mineral horizon thereby creating a relatively thick Ah horizon. Usually many soil organisms are present, but it may also form as a result of the decomposition of dense root networks. Insect droppings and earthworms are usually abundant.

Peatymor: The peatymor humus form is strongly associated with lowland, poorly, or very poorly drained sites. It is sharply delineated from the mineral soil and comprises Of, Om, and/or Oh horizons.

Raw moder: The raw moder humus form is characterized as being transitional between the moder and mor humus forms. It has an L, F, and a thin Hi horizon that is composed of organic granules intermixed with loose mineral grains.

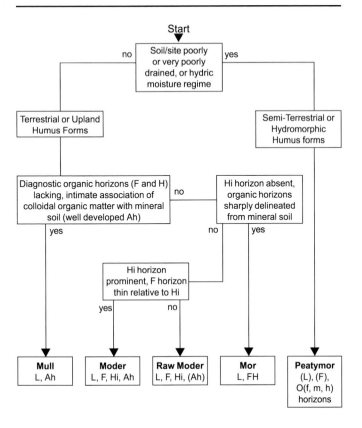

Figure 6.   Key to the humus forms (*modified after* Bernier 1968).

5.6.2   Soil texture

Texture is the relative proportion of sand (2.0 to 0.05 mm), silt (0.05 to 0.002 mm), and clay (<0.002 mm) in a mineral soil sample. Soil texture can be expressed by 14 textural categories as found in a soil textural triangle (Figure 7). In the texture triangle, percent composition of sand and clay are represented along the horizontal and vertical axes, respectively. Percent composition of silt is highest when clay and sand content are low (bottom left corner of texture triangle). For organic soils, texture can be replaced with organic matter decomposition classes (Appendix 1).

To apply the field guide, the texture of each soil horizon is not required. Effective texture and surface texture must be determined. Effective texture for mineral soils is generally defined as the textural class of the finest-textured horizon that occurs 20 to 60 cm below the mineral soil surface and is at least 10 cm thick. The effective texture for organic soils is the dominant organic matter decomposition class (fibric, mesic, or humic) 40 to 80 cm below the surface (Appendix 1).

For mineral soils, surface texture is the dominant texture within the top 20 cm. For organic soils, surface texture is the dominant organic matter decomposition class within the top 40 cm of organic material.

In the absence of laboratory testing, hand texturing is often used to determine a soil sample's textural class (Table 2). The field determination of soil texture is subjective and can only be accomplished consistently with training and experience. Two keys are provided to assist in the field determination of texture (Figures 8 and 9). Although these keys are organized differently and use different techniques, their purposes are the same. The choice of a key is a matter of user preference.

Table 2 provides basic guidelines for applying the soil texture tests in Figures 8 and 9. The moisture content is a critical factor in hand texturing of soils. The soil sample should be moist. If the sample is too dry, apply water until the appropriate moisture content is obtained. No water should be visible but a small amount of moisture may be present on the palm after the soil sample has been tightly squeezed and then released.

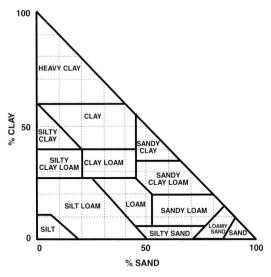

Effective Texture Class	Texture Code	Texture
1	cS	coarse sand
1	mS	medium sand
1	LcS	loamy coarse sand
1	fS	fine sand
1	LmS	loamy medium sand
1	LfS	loamy fine sand
2	SL	sandy loam
2	SiS	silty sand
3	Si	silt
3	SiL	silt loam
3	L	loam
4	SCL	sandy clay loam
4	CL	clay loam
4	SiCL	silty clay loam
4	SC	sandy clay
4	C	clay
4	SiC	silty clay
4	hC	heavy clay

Figure 7.   Soil textural triangle and textural codes and classes.

Table 2.   **Tests used to determine soil texture in the field**

Field test	Description
Moist cast test	Moist soil is compressed by clenching it in the hand. When the hand is released, if the soil holds together (i.e., forms a cast), the strength of the cast is tested by tossing it from hand to hand. The more durable the cast, the more clay in the soil.
Ribbon test	Moist soil is rolled into a long, thin shape (cigarette shape) and then squeezed out between the thumb and forefinger to form the longest and the thinnest ribbon possible. Soils with high silt content will tend to flake rather than ribbon.
Graininess test	To determine "graininess," rub the soil between your fingers. If it feels "grainy," it is an indication that sand is present. If it feels "floury," it indicates that silts are present.
Dry feel test	For soils with >50% sand. Soil is rubbed in the palm of the hand to dry it. Individual sand particles are separated and their sizes estimated. The sand particles are then allowed to fall out of the hand and the amount of remaining finer material (silt and clay) is noted.
Stickiness test	Soil is wetted and compressed between the thumb and forefinger. The degree of stickiness is determined by noting how strongly it adheres to the thumb and forefinger upon release of pressure and by how much it stretches. Stickiness is a measure of clay content in the soil.
Puff test	Done in conjunction with (or instead of) the dry feel test. After the soil is rubbed dry and the sand particles are separated, blow lightly on the palm of the hand and note the proportion of fine floury material that is removed by the puff of breath.
Taste test	Work a small amount of soil between your front teeth. Silt particles are distinguished as fine "grittiness," unlike sand, which is distinguished as individual grains (grainy). Clay is not gritty.
Shine test	Roll a small amount of moderately dry soil into a ball and rub once or twice with a hard smooth object such as a knife or shovel blade. A shine on the ball indicates clay in the soil.

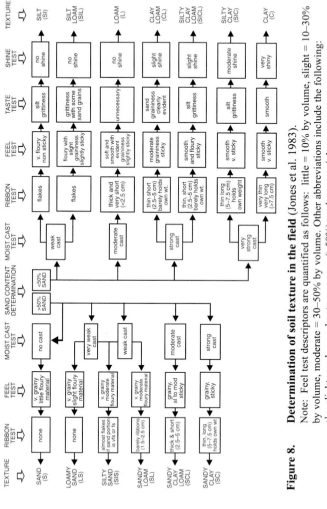

Figure 8.   Determination of soil texture in the field (Jones et al. 1983).

Note: Feel test descriptors are quantified as follows: little = 10% by volume, slight = 10–30% by volume, moderate = 30–50% by volume, very (>50%), and wt = weight. Other abbreviations include the following: sl = slight, mod = moderate, v = very (>50%), and wt = weight.

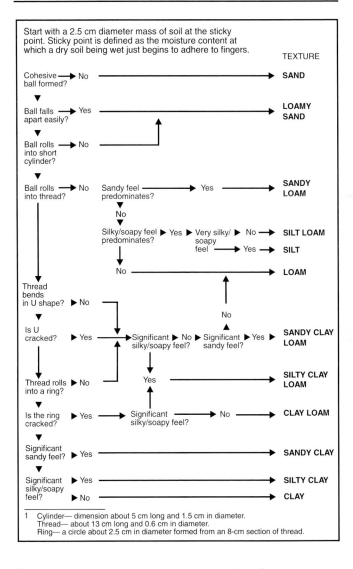

Start with a 2.5 cm diameter mass of soil at the sticky point. Sticky point is defined as the moisture content at which a dry soil being wet just begins to adhere to fingers.

1 Cylinder— dimension about 5 cm long and 1.5 cm in diameter.
Thread— about 13 cm long and 0.6 cm in diameter.
Ring— a circle about 2.5 cm in diameter formed from an 8-cm section of thread.

Figure 9.   Alternative key for hand texturing soil
(after Landon 1988).

5.6.3   Identification of mottles/gley

Mottles and gley are both indicative of permanent or periodic intense reduction of soil minerals through the saturation of soils by water. Gleying occurs on sites or in soil horizons that have wetter moisture conditions than those associated with mottles.

Gleying is a term used to denote the blue, green, bright gray, and sometimes dark bluish-black matrix colors created by anaerobic chemical reactions in water-saturated mineral soils. Due to the anaerobic conditions, minerals in gleyed horizons have been reduced and the soils may emit a rotten egg odor (hydrogen sulfide). Gleying occurs in soils and horizons that have wetter moisture conditions than associated with mottles. To qualify as gleyed, the matrix color must typically have a chroma ≤1, or hues bluer than 10Y based on a Munsell Soil Color Chart (Agriculture Canada Expert Committee on Soil Survey 1987).

Mottles are caused by alternating aerobic/anaerobic soil conditions. Mottles indicate that free water is present in the soil for a significant part of the year. They are indicative of areas of intense oxidation of iron in once-saturated (reducing conditions) mineral soils. Mottles are variable in color but are often rust-colored spots or blotches. Mottles should significantly contrast with the background soil matrix color. There are three levels of contrast between mottle color and the soil matrix color: faint, distinct, and prominent.

Faint: Evident only on close examination. Faint mottles commonly have the same hue as the color to which they are compared and differ by no more than 1 unit of chroma or 2 units of value. Some faint mottles of similar but low chroma and value can differ by 2.5 units of hue.

Distinct: Readily seen, but contrast only moderately with the color to which they are compared. Distinct mottles commonly have the same hue as the color to which they are compared but differ by 2 to 4 units of chroma or 3 to 4 units of value; or differ from the color to which they are compared by 2.5 units of hue but by no more than 1 unit of chroma or 2 units of value.

Prominent: Contrast strongly with the color to which they are compared. Prominent mottles are commonly the most obvious feature in a soil. Prominent mottles that have medium chroma and value commonly differ from the color to which they are compared by at least 5 units of hue if chroma and value are the same; at least 4 units of value or chroma if the hue is the same; or at least 1 unit of chroma or 2 units of value if hue differs by 2.5 units.

5.6.4   Description of parent materials/landforms

Parent materials are the deposits resulting from various landscape creation and destruction processes. Such processes are caused by erosion and/or deposition action related to water, wind, ice, gravity, or *in situ* degradation of the landscape. If you are unfamiliar with the parent materials within a given geographical area, surficial geology maps, physical and ecological land classification maps, and soil surveys can be helpful in providing an overview of the local geomorphology (Strong 1994). The following parent materials occur on the ecosite, ecosite phase, and soil type fact sheets of this field guide.

- Colluvium (C)
- Eolian (E)
- Fluvial (F)
- Lacustrine (L)
- Morainal/Till (M)
- Fluvioeolian (FE)
- Fluviolacustrine (FL)
- Glaciofluvial (GF)

- Glaciolacustrine (GL)
- Rock (R)
- Bog (B)
- Fen (N)
- Marsh (H)
- Swamp (S)
- Undifferentiated Organic (O)

For a description of each parent material refer to Appendix 4, and for a list of soil survey reports relevant to the Mid-Boreal ecoregions of Saskatchewan see Appendix 6.

5.6.5 Determination of drainage

Soil drainage refers to the rate of water removal from the soil in relation to supply. Soil drainage reflects the potential occurrence of soil water in excess of field capacity and the period during which such excess water is present in the plant-root zone. Field capacity is the amount of water that a soil can hold after three days of free drainage (0 to 0.3 bars). Water in excess of field capacity is considered "free water." The prolonged occurrence of free water can result in anaerobic conditions that can create mottles and gleying in the soil.

Soil drainage can be determined by using the information outlined previously in this section to collect the data required to work through the Key to Soil Drainage (Figure 10). The seven drainage classes recognized in Canada are defined below.

Very Rapidly Drained (1): Very rapidly drained soils commonly develop in coarse-textured sands and gravels of fluvial or fluviolacustrine origin. These soils are dry and precipitation is absorbed almost immediately. Groundwater and/or runoff does not influence vegetation growth.

Rapidly Drained (2): Rapidly drained soils commonly develop in medium sands, fine sands, or loamy sands that are generally of fluviolacustrine or eolian origin. Coarse-textured tills also occur in this class. They may contain gravel lenses or be underlain by material of other glacial depositions. Precipitation is almost immediately absorbed, however, at a slightly slower rate than the previous class. Groundwater and/or runoff do not influence vegetation growth. Soil water content seldom exceeds field capacity in any horizon except immediately after water additions. Soils are free from any evidence of gleying or mottling throughout the profile and are often associated with steep slopes.

Well-Drained (3): Well-drained soils have highly variable profile textures and modes of deposition, however, the most common deposit is glacial till, which is frequently of moderately coarse to moderately fine texture. At least one horizon is present that has the ability to significantly restrict water penetration. These soils are water-deficient for short periods and although they may be found on all slope positions, their most common occurrence is from the middle slope to crest positions. Coarse-textured profiles are usually located on lower slopes, such that groundwater and/ or runoff influences vegetation growth and thus differentiates it from the previous two classes. Soil water content does not normally exceed field capacity in any horizon for a significant part of the year. Soils are usually free from mottling in the upper 100 cm, but may be mottled below this depth.

Moderately Well-Drained (4): Moderately well-drained soils contain excess soil water for short periods. At least one horizon is present that has the ability to significantly restrict water penetration. Their characteristic differentiation from the well-drained class is the presence of a few mottles that may occur throughout the profile. However, these soils do not have distinct or prominent mottles above 50 cm.

Imperfectly Drained (5): Excess water for moderately long periods is evident in imperfectly drained soils by the presence of abundant mottling and/or gleying. Soil water is in excess of field capacity for moderately long periods during the year. Soils are distinctly mottled above 50 cm and can be prominently mottled between 50 and 100 cm.

Poorly Drained (6): Poorly drained soils have developed under prolonged saturated or nearly saturated conditions. The mineral substratum, which is gleyed and/or mottled, is usually overlain by a thin layer of peat that may be in various stages of decomposition. Taxonomically, these soils are usually classified within the Gleysolic or Organic Orders (Agriculture Canada Expert Committee on Soil Survey 1987). Generally, they are found on level to undulating topography or in depressional areas.

Very Poorly Drained (7): In very poorly drained soils, free water remains at or within 30 cm of the ground surface most of the year. Prominent mottles or gleying are present within 30 cm of the ground surface, if the organic surface layers are less than 30 cm in thickness. Organic or Gleysolic soils that are more or less continually saturated are the soils in this class.

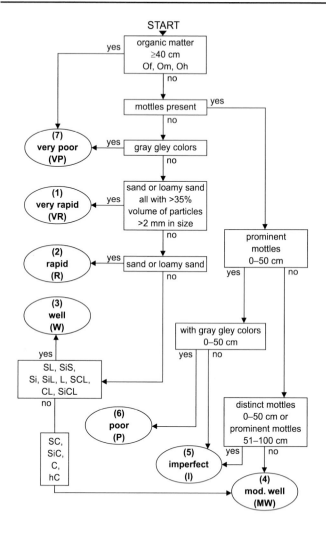

Figure 10. **Key to drainage classes** (Jones et al. 1983). Please refer to the glossary for definitions of prominent and distinct mottles (listed under "Mottles").

5.7 Step 7: Determine important site properties

This section deals with the collection of ecologically important data specific to a particular site. Important site variables such as topographic position, slope, aspect, moisture regime (MR), and nutrient regime (NR) are described briefly below. Subsequent sections provide detailed descriptions about how to evaluate these parameters.

Topographic Position: Topographic position is the relative location of a site on the landscape. The scale is considered a vertical distance between 1 and 300 m and a surface area exceeding 0.5 ha in size. Slope position ranges from crest to depression.

Slope: The slope of a site describes the percentage of vertical rise relative to horizontal distance. Zero degrees as percent slope describes a level site and 45° is equivalent to 100% slope.

Aspect: Aspect describes the orientation of a slope as determined by the points of a compass. It is indicated as slope exposure in degrees or to the north, south, east, west, or points between (e.g., northwest). Aspect, in combination with slope, is important in predicting the amount of solar radiation a site receives. Level topography has no aspect.

Moisture Regime (MR): Moisture regime represents the moisture available for plant growth. It is assessed through an integration of species composition and soil and site characteristics.

Nutrient Regime (NR): Nutrient regime signifies, on a relative scale, the available nutrient supply for plant growth. The determination of nutrient regime requires the integration of many environmental and biotic parameters.

5.7.1 Topographic position

Topographic position is the relative location of a site on the landscape. The scale is considered a vertical distance between 1 and 300 m and a surface area exceeding 0.5 ha in size. Explanations of the topographic position classes are described below and their distribution within the landscape is depicted in Figure 11.

Crest: The generally convex uppermost portion of a hill. It is usually convex in all directions; no distinct aspect.

Upper Slope: The generally convex upper portion of the slope of a hill immediately below the crest. It has a convex surface profile with a specific aspect.

Middle Slope: The area of the slope of a hill between the upper slope and the lower slope, where the slope profile is not generally concave or convex; it has a straight or somewhat sigmoid surface profile with a specific aspect.

Lower Slope: The area toward the base of the hill. It generally has a concave surface profile with a specific aspect.

Toe: The toe is below and adjacent to the lower slope and is clearly demarcated by an abrupt decrease in the slope.

Depression: Any area that is concave in all directions; generally at the foot of a hill or in a generally level area.

Level: An area where the surface profile is generally horizontal with no significant aspect.

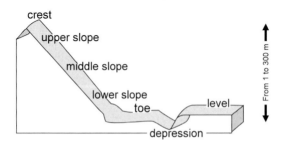

Figure 11.   Schematic depiction of topographic position classes.

5.7.2 Determine moisture regime

Moisture regime (MR) represents the moisture available for plant growth and is an essential component of the ecological classification system. It is assessed through an integration of indicator plant species information and soil and site characteristics. Because moisture regime is a dynamic property that varies throughout the year, it should be evaluated on the basis of the growing season as a whole, not for one specific time. Seasonal variability must be considered when evaluating moisture regime.

The moisture regime classes and potential identifying characteristics are given in Table 3. To use Table 3, go to the appropriate drainage class as determined using information in Section 5.6.5 (Figure 10). Then you can adjust the moisture regime up or down based on topographic position as determined in Section 5.7.1 (Figure 11) and information about the other variables listed in Table 3.

5.7.3 Determine nutrient regime

Nutrient regime (NR) signifies, on a relative scale, the available nutrient supply for plant growth. The determination of nutrient regime requires the integration of several environmental and biotic parameters.

A summary of some of the important factors that affect nutrient regime and are relatively easy to identify are outlined in Figure 12. Using Figure 12, determine the nutrient regime class that most closely represents the conditions on the study site. Humus form and A horizon type will commonly be the most important factors in this determination, unless the site has unusual soil conditions (e.g., very stony or thin), has a high ground water table, or is prone to seepage (Strong 1994). Temporary or permanent seepage, subsurface flow, and flooding usually enrich a site with nutrients.

When two nutrient classes could be equally valid as choices based on your data compared to Figure 12, consider the source of the litter that forms the humus. If the litter originates primarily from deciduous trees and shrubs, select the more nutrient-rich class (Strong 1994). The poorer

Table 3. Characteristics of moisture regime classes

Moisture regime	Description	Primary water source	Topographic position	Effective texture[a]	Soil drainage	Depth to impermeable layer	Surface organic thickness	Slope gradient and aspect
Very xeric (1)	Water removed extremely rapidly in relation to supply; soil is moist for a negligible time after precipitation	Precipitation	Ridge, crest, shedding	Very coarse (gravel-S); abundant coarse fragments (>50%)	Very rapid	Very shallow (<0.5 m)	Very shallow (<3 cm)	Very steep (>70%); southerly aspect
Xeric (2)	Water removed very rapidly in relation to supply; soil is moist for brief periods following precipitation	Precipitation	Ridge, crest, shedding	Coarse to moderately coarse (LS-SL);	Very rapid to rapid	Very shallow (<0.5 m)	Very shallow (<3 cm)	Very steep (>70%); southerly aspect
Subxeric (3)	Water removed rapidly in relation to supply; soil is moist for short periods following precipitation	Precipitation	Upper slope shedding	Coarse to moderately coarse (LS-SL);	Rapid	Shallow (<1.0 m)	Shallow (3–5 cm)	Steep (31–70%); southerly aspect
Submesic (4)	Water removed readily in relation to supply; water available for moderately short periods following precipitation	Precipitation	Upper slope shedding	moderate coarse fragments	Rapid to well	Shallow (<1.0 m)	Moderately shallow (6–9 cm)	Steep (31–70%); southerly to variable aspect
Mesic (5)	Water removed somewhat slowly in relation to supply; soil may remain moist for significant but sometimes short periods of the year; available soil water reflects climatic input	Precipitation in moderate to fine-textured soils and limited seepage in coarse-textured soils	Midslope rolling to flat	Medium (SiL-L) to fine (SCL-C); few coarse fragments	Well to moderately well	Moderately deep (1–2 m)	Moderately deep (10–15 cm)	Slight to moderate (2–30%); variable aspect

Table 3. concluded

Moisture regime	Description	Primary water source	Topographic position	Effective texture[a]	Soil drainage	Depth to impermeable layer	Surface organic thickness	Slope gradient and aspect
Subhygric (6)	Water removed slowly enough to keep the soil wet for a significant part of the growing season; some temporary seepage and possible mottling below 20 cm	Precipitation and seepage	Lower slope receiving	Variable depending on seepage	Moderately well to imperfect	Deep (>2 m)	Moderately deep to deep (10–40 cm)	Slight (2–9%); variable to no aspect
Hygric (7)	Water removed slowly enough to keep the soil wet for most of the growing season; permanent seepage and mottling present; possibly weak gleying	Seepage	Lower slope receiving	Variable depending on seepage	Imperfect to poor	Variable	Deep (16–40 cm)	Slight (2–9%); variable to no aspect
Subhydric (8)	Water removed slowly enough to keep the water table at or near the surface for most of the year; organic and gleyed mineral soils; permanent seepage less than 30 cm below the surface	Seepage or permanent water table	Depression and level receiving	Variable depending on seepage	Poor to very poor	Variable	Very deep (>40 cm)	Slight (2–9%); variable to no aspect
Hydric (9)	Water removed so slowly that the water table is at or above the soil surface all year; organic and gleyed mineral soils	Permanent water table	Depression and level receiving	Variable depending on seepage	Very poor	Variable	Very deep (>40 cm)	Flat (<2%); no aspect

[a] Symbols under effective texture are as follows: L = loam, S = sand, Si = silt, C = clay.

class would be selected if the source were primarily evergreens or ericaceous plants. If the understory vegetation is lush and/or rich in species or has species indicative of rich sites select the richer nutrient class.

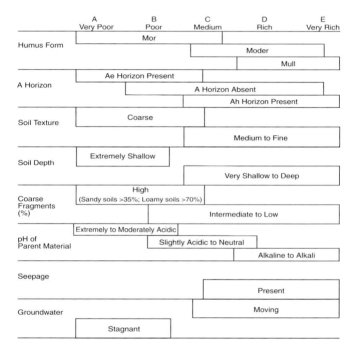

Figure 12. Characteristics of nutrient regime classes
(*modified from* Klinka et al. 1981).

5.8   STEP 8:
Determine the ecoregion, landscape area, ecosite, ecosite phase, and plant community type

First determine the ecoregion and landscape area. This can be done by using the information in Section 2.0 (Figures 1 and 2) and by referring to the Ecoregions of Saskatchewan map (Padbury and Acton 1994). Determining the ecosite, ecosite phase, and plant community type is most accurately done in the field. Always determine the ecological classification of the site before you leave the field.

There are several ways to determine the ecosite, ecosite phase, and plant community type. The first way to assign an ecological classification to a site is to use the field data collected in the previous sections (Sections 5.5 through 5.7) and the key to the ecosite phases in Section 7.0. Once you find a potentially correct ecosite phase, check the soil, site, and vegetational characteristics of the site to make sure it matches the ecosite, ecosite phase, and plant community type descriptions on the fact sheets (Section 8.0). Make sure you identify the site to the plant community type level as this information is crucial for the correct classification of some sites, especially those with limited understory cover (e.g., d3.9 tA-wS/feather moss). While plant community type keys are not provided, you can compare the plant species abundance on the site to the list in the plant community type table on the page facing the ecosite phase fact sheet. Choose the community that best matches the field data.

If the site does not match the ecosite, ecosite phase, or plant community type descriptions, then trace your steps and determine where you may have made a mistake using the key. Also make it a habit to check adjacent ecosites (on the edatope) and their ecosite phases and plant community types to make sure the site does not fit there better. If you have what might be considered a transitional or ecotonal site, then, based on the site conditions, choose the ecosite associated with the best management decisions or the ecosite where the probability of improper management is minimal.

Once you become familiar with the ecological classification system and the ecology of the Mid-Boreal ecoregions of Saskatchewan, you may not want to use the key to classify each site. You must, however, collect the required data (Sections 5.5, 5.6, and 5.7). Once you have collected the required data use moisture and nutrient regime classes to identify the position on the edatopic grid. There may be one or more ecosites that occur within that grid position. All possibilities should be considered. To consider all ecosite choices, you must compare the characteristics of your site with the ecosite, ecosite phase, and plant community type descriptions on the fact sheets for all ecosites that have ellipses that overlap the moisture and nutrient classes of your site on the edatopic grid. Even if you do not require information to the resolution of the plant community type level, each site should be classified to the community type level to increase the probability of correct classification.

If a site has been recently disturbed by fire, logging, or development you may not be able to classify it to the plant community type level. If the canopy is missing and the vegetation and humus layer are disturbed you may only be able to classify the site to the ecosite level. This means that you will be relying more heavily on the moisture and nutrient data because other information on the fact sheets may not match. For each site evaluation, it is crucial that soil, site, and vegetation information are collected with care and accuracy.

5.9   STEP 9: Determine soil type

Once the soil and site data have been collected (Sections 5.6 and 5.7) the key to the soil types can be used to classify the soil type. Additional information about the soil type can be obtained from the soil type fact sheets in Section 9.0. Soil type classification is especially useful when a site has been disturbed.

5.10   STEP 10:
Consult forest management interpretations as required

Interpretations about the ecological units in the classification system have been developed through literature review and expert opinion. They present the user with a general outline of management limitations and opportunities, that, together with the user's knowledge and experience, should be applied in a creative and intelligent manner. Some limitations will change dramatically with time, season of year, economic conditions, existing technology, scale of application, and program objectives (Still and Utzig 1982). The information in this guide should not be construed as a formal recommendation or guideline for resource management, or as a prescription for specific sites.

Management interpretation bars are presented at the bottom of the ecosite and soil type fact sheets. They provide information about the potential limitations and opportunities for the ecological unit. More detailed information about each interpretation and how the levels were defined and derived can be found in Section 12.0. Mensuration and forest inventory information can be found in Section 13.0.

6.0   HOW TO READ THE FACT SHEETS

The field guide contains five types of fact sheets; two for each ecosite, one for each ecosite phase, one for the plant community types within an ecosite phase, and one for each soil type. As well as introducing information on how to interpret the fact sheets, the information on each type of fact sheet will be reviewed briefly.

Variables that represent site and soil characteristics are presented on many of the fact sheets. The classes shown for each variable have a frequency superscript that indicates the proportion of the sample in which each variable class occurred (superscript value $\times$ 10 = percent frequency of occurrence). Because data came from a number of sources and studies over seven years, not all data were collected consistently so data gaps may exist for some variables. Thus, the superscript indicates the frequency of occurrence within the number of sampled plots for which data were collected for that variable. Individual frequency values are rounded to the nearest 10%, consequently total frequency of occurrence may sum to a value other than 100%. The minimum frequency of occurrence a variable had to meet to be included on the fact sheets was 8%.

e.g.,   Parent Material: GF^4, E^3

> 40% (35–44% range) of the plots sampled for the ecological unit occurred on glaciofluvial (GF) parent materials and 30% (25–34% range) occurred on eolian (E) parent material.

e.g.,   Organic Thickness: $(0–5)^8$, $(6–15)^1$

> 80% (75–84% range) of the plots sampled for the ecological unit occurred with an organic matter layer between 0 and 5 cm thick; 10% (8.0–14% range) occurred with an organic matter layer between 6 and 15 cm thick.

Each ecosite is described with two facing fact sheets (Figures 13 and 14). Figure 15 provides a legend to the plant silhouettes and parent material patterns that appear in the ecosite cross section diagrams. This is followed by one page for each ecosite phase that provides biophysical information about the ecosite phase (Figure 16). Facing the ecosite phase page is the plant community type table that displays species cover data for each plant community type belonging to the ecosite phase (Figure 17). The soil type fact sheets present information about the biophysical characteristics of each soil type (Figure 18).

6.1   Ecosite fact sheet 1

There is an identification banner at the top of the first ecosite fact sheet that provides the ecosite letter [1] and name [2] followed by the number of plots [3] (Figure 13). Each ecosite has been given a name that attempts to convey some information about the ecology of the unit. Ecosites are frequently named after plant species that are common or typical of the ecosite. The plant that the ecosite is named after, however, is not present in every plot or stand belonging to that ecosite. A short text description [4] of the ecosite is given under the General Description heading.

The ecosite cross section diagram [5] uses an ecosite phase to depict the position of the ecosite on the landscape, the type of parent materials it commonly occurs on and its topographic relationship to adjacent ecosites. Figure 15 provides a legend to the plant silhouettes and parent material patterns that appear in the ecosite cross section diagrams. The section on successional relationships [6] gives a brief note about the temporal development of the ecosite. In some cases it describes the successional relationship among the ecosite phases and/or plant community types. Below the successional relationships description on the first ecosite fact sheet is a management interpretations bar [7] that provides information about the potential limitations and opportunities for the ecosite. More detailed information about each management interpretation and how the rating levels were defined and derived can be found in Section 12.0. At the bottom left corner of the first page is the page number [8] and at the bottom right is the hierarchical classification level [9] that the page describes.

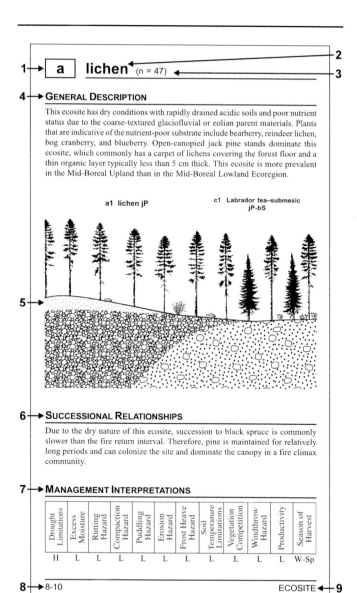

1 → | a | **lichen** (n = 47) ← 2, 3

4 → **GENERAL DESCRIPTION**

This ecosite has dry conditions with rapidly drained acidic soils and poor nutrient status due to the coarse-textured glaciofluvial or eolian parent materials. Plants that are indicative of the nutrient-poor substrate include bearberry, reindeer lichen, bog cranberry, and blueberry. Open-canopied jack pine stands dominate this ecosite, which commonly has a carpet of lichens covering the forest floor and a thin organic layer typically less than 5 cm thick. This ecosite is more prevalent in the Mid-Boreal Upland than in the Mid-Boreal Lowland Ecoregion.

a1 lichen jP

c1 Labrador tea–submesic jP-bS

6 → **SUCCESSIONAL RELATIONSHIPS**

Due to the dry nature of this ecosite, succession to black spruce is commonly slower than the fire return interval. Therefore, pine is maintained for relatively long periods and can colonize the site and dominate the canopy in a fire climax community.

7 → **MANAGEMENT INTERPRETATIONS**

Drought Limitations	Excess Moisture	Rutting Hazard	Compaction Hazard	Puddling Hazard	Erosion Hazard	Frost Heave Hazard	Soil Temperature Limitations	Vegetation Competition	Windthrow Hazard	Productivity	Season of Harvest
H	L	L	L	L	L	L	L	L	L	L	W-Sp

8 → 8-10

ECOSITE ← 9

Figure 13.   An example of the first ecosite fact sheet.

6.2 Ecosite fact sheet 2

The banner at the top of the second ecosite fact sheet identifies the general moisture regime and nutrient regime [1] of the ecosite (Figure 14). At the top right of the page is the letter representing the ecosite [2]. An edatope [3] is used to graphically depict the location of the ecosites with respect to moisture regime and nutrient regime. The ellipse around each ecosite letter represents the approximate distribution of plots belonging to that ecosite. The location of the ecosite on the edatope is shaded. Plant species that are indicative of the ecological conditions of the ecosite are listed under the Indicator Species heading [4]. All sites that belong to a particular ecosite will not have all the indicator species. Some sites will not have any of the indicators so more reliance will have to be placed on the site's moisture and nutrient status for classification. Species-specific site index at 50 years of age at breast height (1.3 m) [5] and mean annual increment [6] are measures of productivity presented to the right of the edatope. The mean site index (SI) is presented in meters followed by the standard error of the mean and the number of sites used to calculate the mean. The arithmetic average of mean annual increment (m^3/ha/yr) (MAI), its standard error, and the number of sites used to calculate the MAI are also presented. For additional information about SI and MAI refer to Section 13.0.

On the second ecosite fact sheet site characteristics [7] and soil characteristics [8] are presented. If you require more information about a site or soil variable and its classes use the table of contents or the glossary (see Section 5.0 and Section 15.0). The ecosite phases [9] that belong to the ecosite are listed, followed by the number of samples defining the ecosite phase. At the bottom left corner of the page is the hierarchical classification level [10] that is described on the page and at the bottom right is the page number [11].

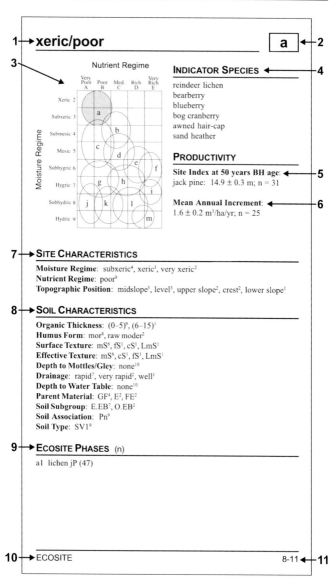

→ xeric/poor

a ←

Nutrient Regime

(Moisture Regime vs Nutrient Regime diagram: Very Poor A, Poor B, Med. C, Rich D, Very Rich E across top; Xeric 2, Subxeric 3, Submesic 4, Mesic 5, Subhygric 6, Hygric 7, Subhydric 8, Hydric 9 down side. Bubbles labelled a, b, c, d, e, f, g, h, i, j, k, l, m.)

INDICATOR SPECIES ←

reindeer lichen
bearberry
blueberry
bog cranberry
awned hair-cap
sand heather

PRODUCTIVITY

Site Index at 50 years BH age: ←
jack pine: 14.9 ± 0.3 m; n = 31

Mean Annual Increment: ←
1.6 ± 0.2 m^3/ha/yr; n = 25

→ SITE CHARACTERISTICS

Moisture Regime: subxeric[4], xeric[3], very xeric[2]
Nutrient Regime: poor[9]
Topographic Position: midslope[3], level[3], upper slope[2], crest[2], lower slope[1]

→ SOIL CHARACTERISTICS

Organic Thickness: (0–5)[9], (6–15)[1]
Humus Form: mor[8], raw moder[2]
Surface Texture: mS[6], fS[1], cS[1], LmS[1]
Effective Texture: mS[6], cS[1], fS[1], LmS[1]
Depth to Mottles/Gley: none[10]
Drainage: rapid[7], very rapid[2], well[1]
Depth to Water Table: none[10]
Parent Material: GF[4], E[2], FE[2]
Soil Subgroup: E.EB[7], O.EB[2]
Soil Association: Pn[9]
Soil Type: SV1[9]

→ ECOSITE PHASES (n)

a1 lichen jP (47)

→ ECOSITE   8-11 ←

Figure 14.   An example of the second ecosite fact sheet.

Trees

Coniferous Softwoods

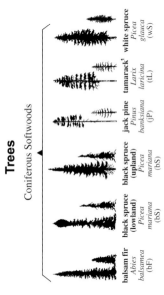

| balsam fir
Abies
balsamea
(bF) | black spruce
(lowland)
Picea
mariana
(bS) | black spruce
(upland)
Picea
mariana
(bS) | jack pine
Pinus
banksiana
(jP) | tamarack[1]
Larix
laricina
(tL) | white spruce
Picea
glauca
(wS) |

Deciduous Hardwoods

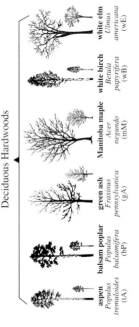

| aspen
Populus
tremuloides
(tA) | balsam poplar
Populus
balsamifera
(bP) | green ash
Fraxinus
pennsylvanica
(gA) | Manitoba maple
Acer
negundo
(mM) | white birch
Betula
papyrifera
(wB) | white elm
Ulmus
americana
(wE) |

Figure 15.   Plant silhouettes and parent material patterns.

[1] Tamarack is a deciduous conifer.

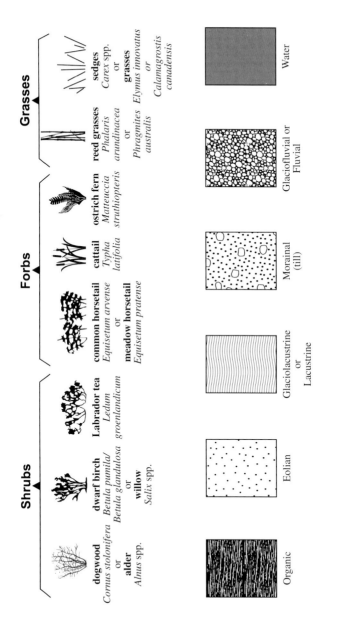

Figure 15. concluded.

6.3 Ecosite phase fact sheet

There is an identification banner at the top of the ecosite phase fact sheet that provides the ecosite phase code [1] and name [2] (Figure 16). Each ecosite phase has been given a name that combines the dominant species in the canopy (phase) with the ecosite name (ecosite). Following the ecosite phase name is the number of plots [3] that were used to define the ecosite phase. An edatope [4] is used to graphically depict the moisture and nutrient grid location of the ecosite to which the ecosite phase belongs. The ellipse around each ecosite represents the approximate distribution of plots belonging to that ecosite and the location of the ecosite is identified with shading. The mean percent cover of plant species that are characteristic of the ecosite phase [5] are presented for each stratum. A plus sign (+) is used to represent species cover if the mean cover for that species was less than 0.5%. Characteristic species are plant species that were either present in a minimum of 70% of the sample plots or had a prominence value of 20 or greater (prominence value = $\sqrt{\% \text{ frequency} \times \% \text{ cover}}$). Scientific names for the species can be found either on the adjacent page in the "Plant Community Types" table or in Appendix 7.

At the top of the second column, site characteristics [6] are presented. Stand age [7] is the mean total age in years of the single dominant tree species for the plots used to define the ecosite phase at the time the ecological data were collected. Following the mean age is the standard error of the mean and then the number of observations used in the calculation. Age data for all ecosite phases are listed in Section 13.0. Richness [8] is the mean vascular plant species richness or the mean number of species of vascular plants followed by the standard error of the mean and the number of samples used in the calculation. Diversity [9] is the mean vascular plant species diversity calculated using the Shannon-Wiener index (Krebs 1989). The standard error of the mean and the number of samples used in the calculation of the Shannon-Wiener index are also presented. Additional information about species richness

1 **a1** **lichen jP** (n = 47) 3

2

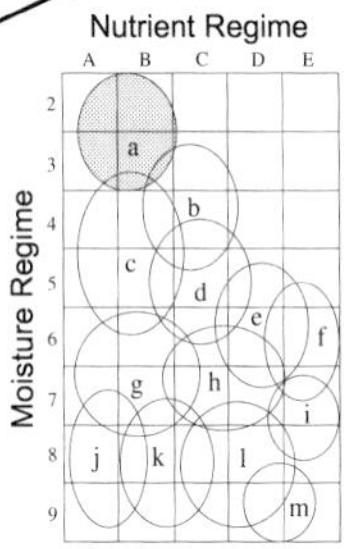

4

5 **CHARACTERISTIC SPECIES**

Tree
[37] jack pine

Shrub
[9] bearberry
[7] blueberry
[4] bog cranberry
[3] green alder*
[1] sand heather*

Forb
[2] wild lily-of-the-valley

Moss
[16] Schreber's moss
[1] cushion moss

Lichen
[28] reindeer lichen

SITE CHARACTERISTICS 6

Moisture Regime: subxeric[4], xeric[3], very xeric[2]
Nutrient Regime: poor[9]
Topographic Position: midslope[3], level[3], upper slope[2], crest[2], lower slope[1]

Stand Age: 69.9 ± 4.0 yrs; n = 34 7
Richness: 16.25 ± 1.05; n = 47 8
Diversity: 1.90 ± 0.11; n = 47 9

SOIL CHARACTERISTICS 10

Organic Thickness: (0–5)[9], (6–15)[1]
Humus Form: mor[8], raw moder[2]
Surface Texture: mS[6], cS[1], fS[1], LmS[1]
Effective Texture: mS[6], fS[1], cS[1], LmS[1]
Depth to Mottles/Gley: none[10]
Drainage: rapid[7], very rapid[2], well[1]
Depth to Water Table: none[10]
Parent Material: GF[4], E[2], FE[2]
Soil Subgroup: E.EB[7], O.EB[2]
Soil Association: Pn[9]
Soil Type: SV1[9]

PLANT COMMUNITY TYPES (n) 11

a1.1 jP/bearberry/lichen (23)
a1.2 jP/blueberry/lichen (14)
a1.3 jP/green alder/lichen (10)

* Species characteristic of the phase but occurring in <70% of the sample plots with a prominence value <20.

12 8-12 ECOSITE PHASE 13

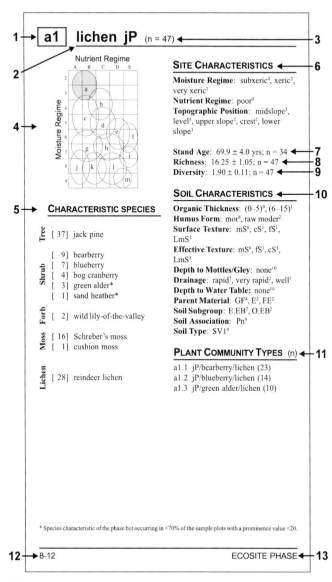

Figure 16. An example of the ecosite phase fact sheet.

and diversity can be found in Section 12.12.1. Following site characteristics, data about soil characteristics [10] are presented. If you require more information about a site or soil variable and its classes, use the table of contents, the glossary, or Section 5.0. The plant community types [11] that occur within the phase are listed with their code (e.g., a1.1) and name (jP/bearberry/lichen) followed by the number of samples used to define the community. At the bottom left corner of the page is the page number [12] and at the bottom right is the hierarchical classification level [13] that the page describes.

6.4   Plant community type table

Facing the ecosite phase fact sheet is the plant community type table (Figure 17) depicting prominent species for each community type for the ecosite phase. The title at the top of the table identifies the ecosite phase that the table describes [1]. Below the title there are codes for the plant community types that are presented for the ecosite phase [2]. A cover value for a plant species was only displayed in the plant community type table if in at least one plant community type it had a prominence value greater than 20 (prominence value = $\sqrt{\% \text{ frequency} \times \% \text{ cover}}$) [3]. The scientific name [4], common name [5], and stratum or layer [6] are provided for each species. A legend for the cover classes is provided under the table [7] and at the bottom of the page is the page number [8].

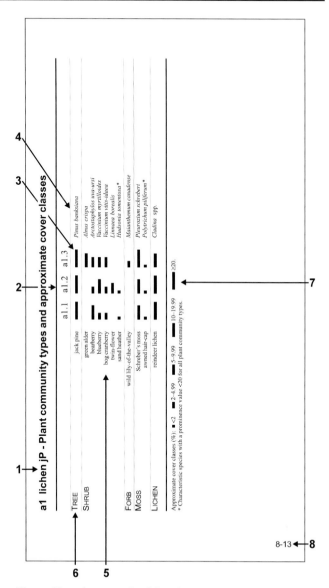

a1 lichen jP - Plant community types and approximate cover classes

		a1.1	a1.2	a1.3	
TREE	jack pine				*Pinus banksiana*
SHRUB	green alder				*Alnus crispa*
	bearberry				*Arctostaphylos uva-ursi*
	blueberry				*Vaccinium myrtilloides*
	bog cranberry				*Vaccinium vitis-idaea*
	twin-flower				*Linnaea borealis*
	sand heather				*Hudsonia tomentosa **
FORB	wild lily-of-the-valley				*Maianthemum canadense*
MOSS	Schreber's moss				*Pleurozium schreberi*
	awned hair-cap				*Polytrichum piliferum **
LICHEN	reindeer lichen				*Cladina* spp.

Approximate cover classes (%): ■ <2   ■ 2–4.99   ■ 5–9.99   ■ 10–19.99   ■ ≥20.
* Characteristic species with a prominence value <20 for all plant community types.

Figure 17.   An example of the plant community type table.

6.5   Soil type fact sheet

There is an identification banner at the top of the soil type fact sheet that provides the soil type code [1] and name [2] including sample size [3] (Figure 18). Below the soil type name is the general description [4]. Next to the general description is the soil textural triangle [5] with the actual percent frequency of occurrence of the effective texture represented by tones of shading of the texture classes on the triangle. Soil [6] and site [7] characteristics are presented as the frequency of occurrence of each class for each variable. If you require more information about a site or soil variable and its classes use the table of contents, the glossary, or Section 5.0.

Next to the site characteristics is a typical soil profile [8] for the soil type. Information about the soil horizon designations can be found in Appendix 2 and *The Canadian system of soil classification* (Agriculture Canada Expert Committee on Soil Survey 1987). Where a horizon code is in brackets it means that the horizon characteristic occurred in a small but significant number of sample plots. The variability in the depth of the various horizons is depicted by the change in the thickness of the horizon. If the horizon does not extend across the profile picture, then it occurs in approximately the number of sample sites proportional to the distance the horizon extends across the profile diagram.

Under the Associated Ecosites heading [9] the frequency of occurrence of each ecosite within the soil type is displayed. There is a comments section [10] near the bottom of each fact sheet where miscellaneous notes and observations are provided. The management interpretations bar [11] provides information about the potential limitations and opportunities for the soil type. Additional information about each management interpretation and how the rating levels were defined and derived can be found in Section 12.0. At the bottom of the page is the page number [12] and the classification level [13] that the page describes.

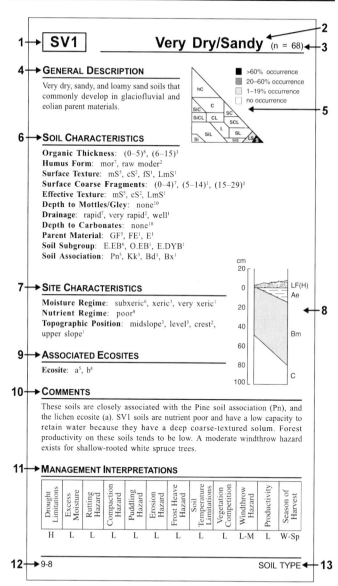

1 → **SV1**   **2** → **Very Dry/Sandy** **3** → (n = 68)

4 → ## GENERAL DESCRIPTION

Very dry, sandy, and loamy sand soils that commonly develop in glaciofluvial and eolian parent materials.

5 →
- ■ >60% occurrence
- ▨ 20–60% occurrence
- ▢ 1–19% occurrence
- □ no occurrence

6 → ## SOIL CHARACTERISTICS

Organic Thickness: $(0–5)^8$, $(6–15)^3$
Humus Form: mor[7], raw moder[2]
Surface Texture: mS[5], cS[2], fS[1], LmS[1]
Surface Coarse Fragments: $(0–4)^7$, $(5–14)^1$, $(15–29)^1$
Effective Texture: mS[5], cS[2], LmS[1]
Depth to Mottles/Gley: none[10]
Drainage: rapid[7], very rapid[2], well[1]
Depth to Carbonates: none[10]
Parent Material: GF[5], FE[1], E[1]
Soil Subgroup: E.EB[6], O.EB[1], E.DYB[1]
Soil Association: Pn[5], Kk[3], Bd[1], Bx[1]

7 → ## SITE CHARACTERISTICS

Moisture Regime: subxeric[6], xeric[3], very xeric[1]
Nutrient Regime: poor[9]
Topographic Position: midslope[3], level[3], crest[2], upper slope[1]

8 → (soil profile diagram: cm 20, 0, 20, 40, 60, 80, 100; LF(H), Ae, Bm, C)

9 → ## ASSOCIATED ECOSITES

Ecosite: a[5], b[4]

10 → ## COMMENTS

These soils are closely associated with the Pine soil association (Pn), and the lichen ecosite (a). SV1 soils are nutrient poor and have a low capacity to retain water because they have a deep coarse-textured solum. Forest productivity on these soils tends to be low. A moderate windthrow hazard exists for shallow-rooted white spruce trees.

11 → ## MANAGEMENT INTERPRETATIONS

Drought Limitations	Excess Moisture	Rutting Hazard	Compaction Hazard	Puddling Hazard	Erosion Hazard	Frost Heave Hazard	Soil Temperature Limitations	Vegetation Competition	Windthrow Hazard	Productivity	Season of Harvest
H	L	L	L	L	L	L	L	L	L-M	L	W-Sp

12 → 9-8   SOIL TYPE ← **13**

Figure 18.   An example of the soil type fact sheet.

7.0 KEY TO THE ECOSITES AND ECOSITE PHASES

This section provides keys to the ecosite phases that enable the user to identify and locate in the guide a particular ecosite phase. It also contains a synopsis of the relationships among the ecosites in the form of several cluster analysis dendrograms (Figures 19 and 20).

The ecological relationships among the upland ecosites and among the lowland ecosites have been presented graphically with the use of several cluster analysis dendrograms (Figures 19 and 20). These diagrams depict the mathematical relationship among the ecosites based on a hierarchical agglomerative average linkage cluster analysis of mean plant species cover data and mean moisture regime, drainage, and nutrient regime for each ecosite (SAS Institute Inc. 1990). Only plant species that were present in a minimum of 70% of the sample plots or had a prominence value of 20 or greater (prominence value = $\sqrt{\% \text{ frequency} \times \% \text{ cover}}$) for at least one plant community type were used in the cluster analysis.

Stepwise canonical discriminant analysis was used to determine which variables were statistically significant in their "discrimination power" between each set of two groups of ecosites defined at each hierarchical level of the hierarchical cluster analysis (SAS Institute Inc. 1990). The same variables were used in the discriminant analysis as were used in the cluster analysis (see above paragraph). Variables that were significantly different (F-test) between any two cluster groups are labeled on Figures 19 and 20 with the appropriate asterisks (*). On the dendrograms, the variable is placed on the side of the "distance line" with the group to which it has an affinity. For example, the percent cover of ostrich fern was significantly different (Prob > F <0.001) between the ostrich fern ecosite (f) and the other medium to nutrient-rich ecosites to the left of it on Figure 19. Therefore, the variable name "ostrich fern" was placed on the side of the "distance line" closest to the ostrich fern ecosite (f), which has high ostrich fern cover.

Although the ecosite dendrograms provide information about the relationship among the ecosites based on plant species cover, moisture regime, nutrient regime, and drainage, they are generalized and should not be used as keys, to the ecosites. They should be used as one of many tools that assist in classifying ecological units and understanding

---

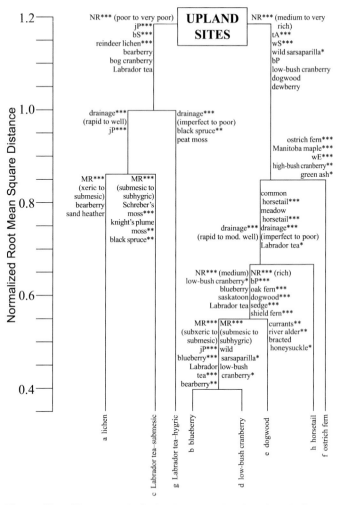

Figure 19. Cluster analysis dendrogram showing the relation-ship among upland ecosites.

Variables listed beside a line are characteristic of the ecosites to which the line leads. The variable is placed on the side of the "distance line" with the ecosite(s) to which it has an affinity. The most significant variables that differentiate ecosites are identified based on canonical discriminant analysis. The level of significance associated with a variable is indicated as follows: *** < 0.001, ** < 0.01, and * < 0.05.

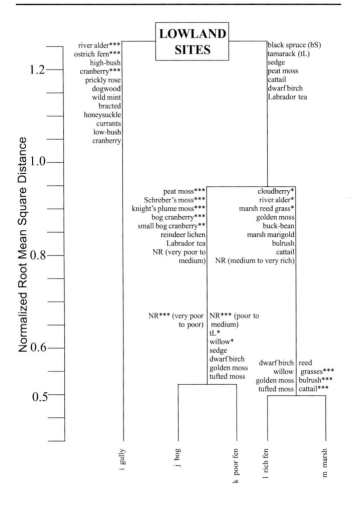

Figure 20. Cluster analysis dendrogram showing the relationship among lowland ecosites.

Variables listed beside a line are characteristic of the ecosites to which the line leads. The variable is placed on the side of the "distance line" with the ecosite(s) to which it has an affinity. The most significant variables that differentiate ecosites are identified based on canonical discriminant analysis. The level of significance associated with a variable is indicated as follows: *** < 0.001, ** < 0.01, and * < 0.05.

ecological relationships. Graphic keys to the ecosite phases are presented in Figures 21 through 25 while text keys are found in Section 7.1.

To use the keys to the ecosite phases, the user must collect data from the field site (Sections 5.5, 5.6, and 5.7). Several terms used in the keys such as "common" and "generally" may seem ambiguous. The highly dynamic and variable nature of ecosystems makes it difficult, in some cases, to define the specific quantities and characteristics that differentiate ecological groups. This difficulty is further intensified because of compensating ecological factors. This means that one factor or group of factors may compensate for a deficiency in other factors. Certain ecological factors existing on the site may interact to produce the same effects as other factors or similar effects as different magnitudes of the same factors. The keys, therefore, attempt to direct the user to the proper ecosite phase by describing general conditions without rigid rules. Where actual quantities are used they should not be too rigidly interpreted. Dominant species refer to the species with the highest percent cover, while a codominant species is one of several species that have relatively high covers.

Once a potentially correct classification is determined, always check the soil, site, and vegetation characteristics of the site to ensure that they match the ecosite, ecosite phase, and plant community type descriptions on the fact sheets (Section 8.0). Be sure to classify the site to the plant community type level as this information is crucial to the correct classification of some sites, especially those with limited understory cover (e.g., d3.9 tA-wS/feather moss). Compare the plant species abundance at your site to the list in the plant community type table on the page facing the ecosite phase fact sheet and choose the community that best matches the field data.

If the site does not match the descriptions at the ecosite, ecosite phase, or plant community type levels, then trace your steps and determine where you may have made a mistake using the key. Also make it a habit to always check the ecosites and associated ecosite phases and plant community types that are adjacent on the edatope. Your site may fit better into one of these groups. If you have what might be considered an ecotonal site, then, based on the site conditions, choose the ecosite that would be associated with the best management decisions or the ecosite where the probability of improper management would be minimized.

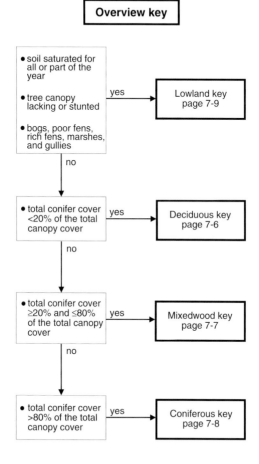

Overview key

- soil saturated for all or part of the year
- tree canopy lacking or stunted
- bogs, poor fens, rich fens, marshes, and gullies

→ yes → Lowland key page 7-9

↓ no

- total conifer cover <20% of the total canopy cover

→ yes → Deciduous key page 7-6

↓ no

- total conifer cover ≥20% and ≤80% of the total canopy cover

→ yes → Mixedwood key page 7-7

↓ no

- total conifer cover >80% of the total canopy cover

→ yes → Coniferous key page 7-8

Figure 21. **Graphic overview key to the ecosite phases.**
Tamarack is a deciduous conifer tree and has been included in the Lowland key.

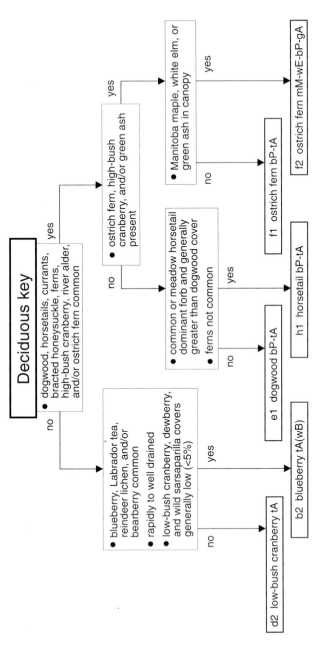

Figure 22. Graphic key to the deciduous tree dominated ecosite phases.

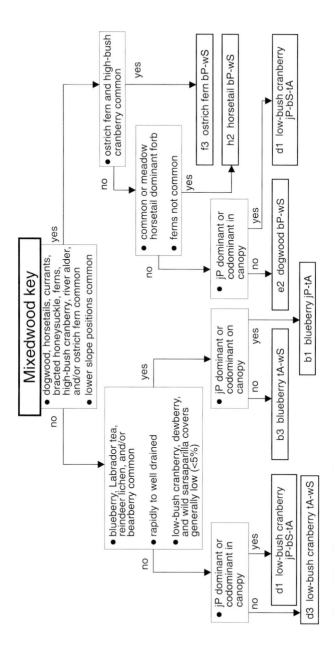

Figure 23. Graphic key to the mixedwood ecosite phases.

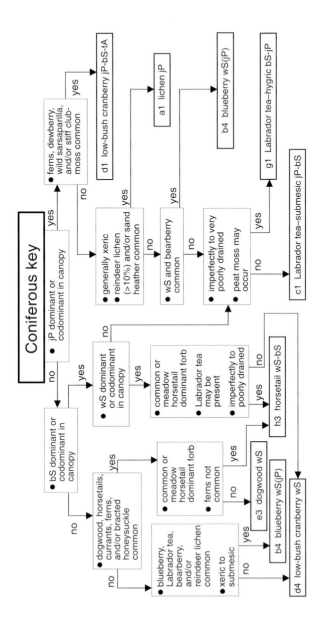

Figure 24. Graphic key to the conifer tree dominated ecosite phases.

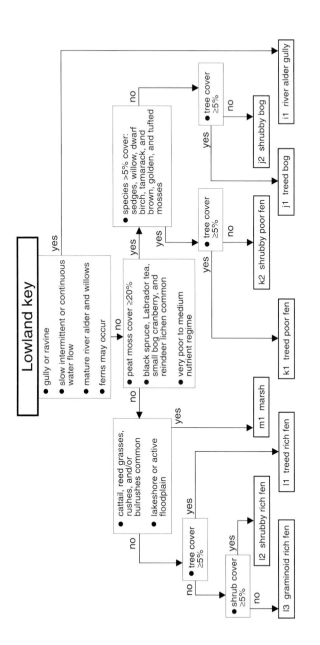

Figure 25. Graphic key to the lowland ecosite phases.

7.1   Overview key

Ia. Soil saturated for part or all of the year; tree canopy lacking, or if present, bS and/or tL are usually stunted due to elevated water table levels; includes bogs, poor fens, rich fens, marshes, and gullies .... **Lowland key.**

Ib. Soil not saturated for extended periods; tree canopy exists (upland sites) .... **II.**

IIa. Total conifer cover <20% of the total canopy cover .... **Deciduous key.**

IIb. Total conifer cover ≥20% of the total canopy cover .... **III.**

IIIa. Total conifer cover ≥20% and ≤80% of the total canopy cover .... **Mixedwood key.**

IIIb. Total conifer cover >80% of the total canopy cover .... **Coniferous key.**

7.1.1   Deciduous key

1a. Any of dogwood, common or meadow horsetail, ferns, river alder, currants, high-bush cranberry, ostrich fern, and bracted honeysuckle are dominant or abundant; soil generally moderately well to poorly drained; site mesic to hygric; lower slope position .... **key lead 2.**

1b. Any of dogwood, common or meadow horsetail, ferns, river alder, currants, high-bush cranberry, ostrich fern, and bracted honeysuckle are not dominant or abundant; soil rapidly to well drained; site subxeric to mesic; upper to midslope position .... **key lead 5.**

2a. Ostrich fern, high-bush cranberry, and/or green ash present .... **key lead 3.**

2b. Ostrich fern, high-bush cranberry, and green ash not present .... **key lead 4.**

3a. Manitoba maple, white elm, or green ash not common
   = **f1 ostrich fern bP-tA** .... p. 8-54

   > **f1.1 bP-tA/dogwood/ostrich fern**
   > **f1.2 bP-tA/mountain maple/ostrich fern**

3b. Manitoba maple, white elm, and/or green ash dominant or codominant in canopy
   = **f2 ostrich fern mM-wE-bP-gA** .... p. 8-56

   > **f2.1 mM-wE-bP-gA/pin cherry–saskatoon/ ostrich fern**
   > **f2.2 mM-wE-bP-gA/Manitoba maple/ ostrich fern**
   > **f2.3 mM-wE-bP-gA/high-bush cranberry– green ash/ostrich fern**

4a. Horsetail cover generally > dogwood cover; common or meadow horsetail dominant forb; ferns not common
   = **h1 horsetail bP-tA** .... p. 8-66

   > **h1.1 bP-tA/horsetail**

4b. Horsetail cover generally ≤ dogwood cover; common or meadow horsetail not dominant forb; ferns may be present
   = **e1 dogwood bP-tA** .... p. 8-44

   > **e1.1 bP-tA/dogwood/fern**
   > **e1.2 bP-tA/bracted honeysuckle/fern**
   > **e1.3 bP-tA/river alder–green alder/fern**
   > **e1.4 bP-tA/alder-leaved buckthorn**
   > **e1.5 bP-tA/mountain maple**

5a. Blueberry, Labrador tea, bearberry, and reindeer lichen common; low-bush cranberry, dewberry, and wild sarsaparilla generally low in cover (<5%); site usually on upper slope positions and/or coarse-textured parent material

= **b2 blueberry tA(wB)** .... p. 8-18

    b2.1 tA(wB)/blueberry–bearberry
    b2.2 tA(wB)/blueberry–green alder
    b2.3 tA(wB)/blueberry–Labrador tea

5b. Blueberry, Labrador tea, bearberry, and reindeer lichen not common; low-bush cranberry, dewberry, and wild sarsaparilla generally high in cover (>5%); site well to moderately well drained

= **d2 low-bush cranberry tA** .... p. 8-32

    d2.1 tA/pin cherry–saskatoon
    d2.2 tA/beaked hazelnut
    d2.3 tA/green alder
    d2.4 tA/low-bush cranberry–prickly rose
    d2.5 tA/willow
    d2.6 tA/bush honeysuckle
    d2.7 tA/mountain maple
    d2.8 tA/forb

7.1.2 Mixedwood key

1a. Any of dogwood, common or meadow horsetail, ferns, river alder, currants, high-bush cranberry, ostrich fern, and bracted honeysuckle are dominant or abundant; soil generally moderately well to poorly drained; site mesic to hygric; lower slope position .... **key lead 2.**

1b. Any of dogwood, common or meadow horsetail, ferns, river alder, currants, high-bush cranberry, ostrich fern, and bracted honeysuckle are not dominant or abundant; soil generally rapidly to well drained; site subxeric to mesic; upper to midslope position .... **key lead 5.**

2a. Ostrich fern and/or high-bush cranberry present
= **f3   ostrich fern bP-wS** .... p. 8-58

      f3.1   bP-wS/dogwood/ostrich fern

2b. Ostrich fern and high-bush cranberry not present .... **key lead 3.**

3a. Horsetail cover generally > dogwood cover, common
or meadow horsetail dominant forb; ferns not common
= **h2   horsetail bP-wS** .... p. 8-68

      h2.1   bP-wS/horsetail

3b. Horsetail cover generally ≤ dogwood cover, common
or meadow horsetail not dominant forb; ferns may be
present .... **key lead 4.**

4a. jP dominant or codominant in canopy
= **d1   low-bush cranberry jP-bS-tA** .... p. 8-30

      d1.1   jP-bS-tA/stiff club-moss–fern

4b. jP not common in canopy
= **e2   dogwood bP-wS** .... p. 8-46

      e2.1   bP-wS/dogwood/fern
      e2.2   bP-wS/bracted honeysuckle/fern
      e2.3   bP-wS/river alder–green alder/fern
      e2.4   bP-wS/bush honeysuckle
      e2.5   bP-wS/mountain maple
      e2.6   bP-wS/balsam fir/fern
      e2.7   bP-wS/fern/feather moss

5a. Blueberry, Labrador tea, bearberry, and reindeer lichen common; low-bush cranberry, dewberry, and wild sarsaparilla tend to be low in cover (<5%); site rapidly to well drained, usually upper slope position and/or coarse-textured parent material .... **key lead 6.**

5b. Blueberry, Labrador tea, bearberry, and reindeer lichen not common; low-bush cranberry, dewberry, and wild sarsaparilla covers tend to be high (>5%); site is well to moderately well drained .... **key lead 7.**

6a. jP is codominant with deciduous component of stand
  = **b1 blueberry jP-tA** .... p. 8-16

> **b1.1 jP-tA/blueberry–bearberry**
> **b1.2 jP-tA/blueberry–green alder**

6b. jP is not codominant with deciduous component of stand
  = **b3 blueberry tA-wS** .... p. 8-20

> **b3.1 tA-wS/blueberry–bearberry**
> **b3.2 tA-wS/blueberry–green alder**
> **b3.3 tA-wS/blueberry–Labrador tea**

7a. jP is dominant or codominant; ferns may be present
  = **d1 low-bush cranberry jP-bS-tA** .... p. 8-30

> **d1.1 jP-bS-tA/stiff club-moss–fern**

7b. jP is not dominant or codominant; ferns not present

        d3.1 tA-wS/pin cherry–saskatoon

        d3.2 tA-wS/beaked hazelnut

        d3.3 tA-wS/green alder

        d3.4 tA-wS/low-bush cranberry–prickly rose

        d3.5 tA-wS/bush honeysuckle

        d3.6 tA-wS/mountain maple

        d3.7 tA-wS/forb

        d3.8 tA-wS/balsam fir/feather moss

        d3.9 tA-wS/feather moss

7.1.3   Coniferous key

1a. jP dominates or codominates the canopy .... **key lead 2.**

1b. jP does not dominate or codominate the canopy .... **key lead 6.**

2a. Ferns, dewberry, wild sarsaparilla, and/or stiff club-
    moss common

        d1.1 jP-bS-tA/stiff club-moss–fern

2b. Ferns, dewberry, wild sarsaparilla, and stiff club-moss
    not common .... **key lead 3.**

3a. Reindeer lichen cover high (>10%) and sand heather
    may be common; site xeric to subxeric; Labrador tea
    cover usually low (<5%)

        a1.1 jP/bearberry/lichen

        a1.2 jP/blueberry/lichen

        a1.3 jP/green alder/lichen

3b. Reindeer lichen cover low to moderate (<10%) and sand
    heather not common; site subxeric to hygric; Labrador
    tea cover variable .... **key lead 4.**

4a. wS dominant or codominant; bS generally not present;
    Labrador tea generally absent, but if present, with low
    cover (<5%); bearberry generally present
    = **b4 blueberry wS(jP)** .... p. 8-22

        b4.1 wS(jP)/blueberry–bearberry
        b4.2 wS(jP)/blueberry–green alder

4b. wS not dominant or codominant; bS may be present;
    Labrador tea may be present; bearberry generally
    absent, but if present, with low cover (<2%) .... **key lead 5.**

5a. Site is imperfectly to very poorly drained; peat moss
    may be present
    = **g1 Labrador tea–hygric bS-jP** .... p. 8-62

        g1.1 bS-jP/Labrador tea/feather moss
        g1.2 bS-jP/feather moss

5b. Site is rapidly to moderately well drained; peat moss
    not present
    = **c1 Labrador tea–submesic jP-bS** .... p. 8-26

        c1.1 jP-bS/Labrador tea/feather moss
        c1.2 jP-bS/green alder/feather moss
        c1.3 jP-bS/feather moss

6a. bS dominates or codominates the canopy .... **key lead 7.**
6b. bS does not dominate or codominate the canopy .... **key lead 9.**

7a. wS is dominant or codominant in the canopy .... **key lead 8.**
7b. wS is not dominant or codominant in the canopy .... **key lead 5.**

8a.　Site imperfectly to poorly drained; common or meadow
　　　horsetail usually abundant (>8%); Labrador tea may
　　　be present
　　　= **h3　horsetail wS-bS** .... p. 8-70

　　　　　　h3.1　wS-bS/horsetail
　　　　　　h3.2　wS-bS/Labrador tea/horsetail

8b.　Site well to moderately well drained; common and
　　　meadow horsetails not abundant; Labrador tea usually
　　　not present
　　　= **d4　low-bush cranberry wS** .... p. 8-40

　　　　　　d4.1　wS/green alder
　　　　　　d4.2　wS/balsam fir/feather moss
　　　　　　d4.3　wS/feather moss

9a.　Ferns, common or meadow horsetail, dogwood, currants,
　　　and/or bracted honeysuckle are generally abundant ...... **key lead 10.**

9b.　Ferns, common or meadow horsetail, dogwood,
　　　currants, and/or bracted honeysuckle are generally not
　　　abundant .... **key lead 11.**

10a. Common and/or meadow horsetail dominant forb; ferns
　　　generally not present
　　　= **h3　horsetail wS-bS** .... p. 8-70

　　　　　　h3.1　wS-bS/horsetail
　　　　　　h3.2　wS-bS/Labrador tea/horsetail

10b. Common and/or meadow horsetail not dominant forb;
　　　ferns may be present
　　　= **e3　dogwood wS** .... p. 8-50

　　　　　　e3.1　wS/river alder/fern
　　　　　　e3.2　wS/balsam fir/fern
　　　　　　e3.3　wS/fern/feather moss

11a. Blueberry, Labrador tea, bearberry, and reindeer lichen
common; low-bush cranberry, dewberry, and wild
sarsaparilla generally low in cover (<5%); site subxeric
to submesic

= **b4 blueberry wS(jP)** .... p. 8-22

   b4.1 wS(jP)/blueberry–bearberry
   b4.2 wS(jP)/blueberry–green alder

11b. Blueberry, Labrador tea, bearberry, and reindeer lichen
not common; low-bush cranberry, dewberry, and wild
sarsaparilla generally high in cover (>5%); site
submesic to subhygric

= **d4 low-bush cranberry wS** .... p. 8-40

   d4.1 wS/green alder
   d4.2 wS/balsam fir/feather moss
   d4.3 wS/feather moss

7.1.4   Lowland key

1a. Site is in a gully or ravine surrounded on either side by
upland sites; slow intermittent or continuous water flow
through the site; mature, large river alder, and/or willow
usually forming a dense thicket; ferns usually present

= **i1 river alder gully** .... p. 8-74

   i1.1 river alder/ostrich fern

1b. Site is not as described in 1a; site level or of
low gradient .... **key lead 2.**

2a. Peat moss generally ≥20% cover; black spruce (trees and
shrubs), Labrador tea, bog cranberry, cloudberry, small
bog cranberry, and/or reindeer lichen common; site has a
very poor to medium nutrient cycling regime .... **key lead 3.**

2b.   Peat moss generally <20% cover; black spruce (trees and
      shrubs), Labrador tea, bog cranberry, cloudberry, small
      bog cranberry, and reindeer lichen not common; site has
      a medium to very good nutrient cycling regime .... **key lead 6.**

3a.   Any or all of the following species are >5% cover:
      sedges, willow, dwarf birch, and brown, golden, and
      tufted mosses; tamarack may occur as a tree or
      shrub .... **key lead 4.**

3b.   Any or all of the following species are ≤5% cover:
      sedges, willow, dwarf birch, and brown, golden, and
      tufted mosses; tamarack does not frequently occur as
      either a tree or shrub .... **key lead 5.**

4a.   Tree cover ≥5% and tree dbh >7.0 cm
      = **k1   treed poor fen** .... p. 8-84

            k1.1   bS-tL/dwarf birch/sedge/peat moss

4b.   Tree cover, if present <5% and dbh ≤7.0 cm
      = **k2   shrubby poor fen** .... p. 8-86

            **k2.1   black spruce–tamarack–dwarf birch/
                     sedge/peat moss**

5a.   Tree cover ≥5% and tree dbh >7.0 cm
      = **j1   treed bog** .... p. 8-78

            j1.1   bS/Labrador tea/cloudberry/peat moss

5b. Tree cover, if present, <5% and dbh ≤7.0 cm

= **j2 shrubby bog** .... p. 8-80

      j2.1 black spruce–Labrador tea/
             cloudberry/peat moss

6a. Emergent vegetation such as cattail, reed grass, bulrushes, and rushes are dominant; site is usually along a lakeshore or an active flood plain; sedge organic material variable (wet mineral and organic soils)

= **m1 marsh** .... p. 8-98

      m1.1 cattail marsh
      m1.2 reed grass marsh
      m1.3 bulrush marsh

6b. Emergent vegetation not common; site is an open peatland or in a bowl depression; sedge organic material usually high (organic soils) .... **key lead 7.**

7a. Tree cover ≥5% and tree dbh >7.0 cm

= **l1 treed rich fen** .... p. 8-90

      l1.1 tL/dwarf birch/sedge/brown moss

7b. Tree cover, if present <5% and dbh ≤7.0 cm .... **key lead 8.**

8a. Shrub layer ≥5% cover

= **l2 shrubby rich fen** .... p. 8-92

      l2.1 dwarf birch/sedge/golden moss
      l2.2 willow/sedge/golden moss
      l2.3 willow/marsh reed grass

8b. Shrub layer <5% cover

= **l3 graminoid rich fen** .... p. 8-94

      l3.1 sedge fen
      l3.2 marsh reed grass fen

8.0 ECOSITE, ECOSITE PHASE, AND PLANT COMMUNITY TYPE FACT SHEETS

The fact sheets in this section provide detailed information about the ecology of the ecological units (ecosite, ecosite phase, and plant community type) of the classification system.

Refer to Section 6.0 for information about how to read the fact sheets. The location of the ecosites on the edatope are shown with an ellipse around each ecosite. The ellipse represents the approximate distribution of plots belonging to that ecosite (Figure 26). Figure 27 shows all the ecological units for each level of the hierarchical classification system including their codes and names. Figure 28 depicts the distribution of the ecosites on the landscape. A key to the symbols used on the cross section diagrams can be found in Figure 15.

To correctly classify a site, field data must be collected (Sections 5.5 to 5.7). These data can be used with the keys in Section 7.0 to assign the field sites to the appropriate ecological units. Once the potentially correct classification is determined, the field data must be compared to the appropriate fact sheets.

Each ecosite is described with two facing fact sheets. This is followed by an ecosite phase fact sheet that provides biophysical information about the ecosite phase. Facing the ecosite phase fact sheet is the plant community type table that displays species cover data for each plant community type belonging to the ecosite phase.

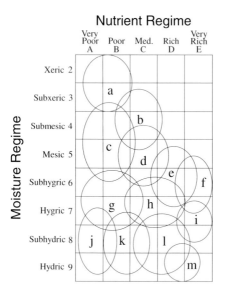

Nutrient Regime

Ecosites:

a = **lichen**
  xeric/poor

b = **blueberry**
  submesic/medium

c = **Labrador tea–submesic**
  submesic/poor

d = **low-bush cranberry**
  mesic/medium

e = **dogwood**
  subhygric/rich

f = **ostrich fern**
  subhygric/very rich

g = **Labrador tea–hygric**
  hygric/poor

h = **horsetail**
  hygric/rich

i = **gully**
  hygric/very rich

j = **bog**
  subhydric/very poor

k = **poor fen**
  subhydric/medium

l = **rich fen**
  subhydric/rich

m = **marsh**
  hydric/rich

**Figure 26.   Edatope (moisture/nutrient grid) showing the
    locations of ecosites.**

Figure 27. Ecological units of the classification system.

ECOSITE	ECOSITE PHASE	PLANT COMMUNITY TYPE
a lichen (xeric/poor)	a1 lichen jP	a1.1 jP/bearberry/lichen
		a1.2 jP/blueberry/lichen
		a1.3 jP/green alder/lichen
b blueberry (submesic/medium)	b1 blueberry jP–tA	b1.1 jP–tA/blueberry–bearberry
		b1.2 jP–tA/blueberry–green alder
	b2 blueberry tA(wB)	b2.1 tA(wB)/blueberry–bearberry
		b2.2 tA(wB)/blueberry–green alder
		b2.3 tA(wB)/blueberry–Labrador tea
	b3 blueberry tA–wS	b3.1 tA–wS/blueberry–bearberry
		b3.2 tA–wS/blueberry–green alder
		b3.3 tA–wS/blueberry–Labrador tea
	b4 blueberry wS(jP)	b4.1 wS(jP)/blueberry–bearberry
		b4.2 wS(jP)/blueberry–green alder
c Labrador tea–submesic (submesic/poor)	c1 Labrador tea–submesic jP–bS	c1.1 jP–bS/Labrador tea/feather moss
		c1.2 jP–bS/green alder/feather moss
		c1.3 jP–bS/feather moss

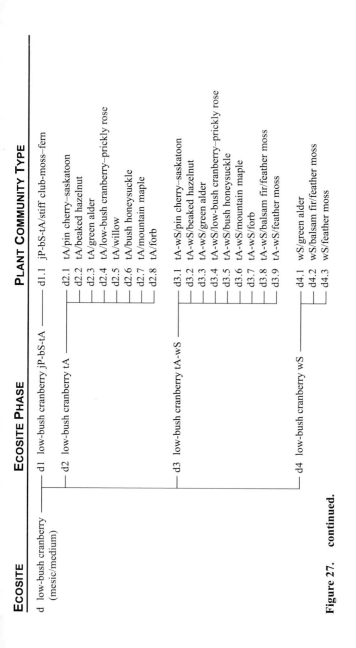

ECOSITE	ECOSITE PHASE	PLANT COMMUNITY TYPE
d low-bush cranberry (mesic/medium)	d1 low-bush cranberry jP-bS-tA	d1.1 jP-bS-tA/stiff club-moss-fern
	d2 low-bush cranberry tA	d2.1 tA/pin cherry–saskatoon
		d2.2 tA/beaked hazelnut
		d2.3 tA/green alder
		d2.4 tA/low-bush cranberry–prickly rose
		d2.5 tA/willow
		d2.6 tA/bush honeysuckle
		d2.7 tA/mountain maple
		d2.8 tA/forb
	d3 low-bush cranberry tA-wS	d3.1 tA-wS/pin cherry–saskatoon
		d3.2 tA-wS/beaked hazelnut
		d3.3 tA-wS/green alder
		d3.4 tA-wS/low-bush cranberry–prickly rose
		d3.5 tA-wS/bush honeysuckle
		d3.6 tA-wS/mountain maple
		d3.7 tA-wS/forb
		d3.8 tA-wS/balsam fir/feather moss
		d3.9 tA-wS/feather moss
	d4 low-bush cranberry wS	d4.1 wS/green alder
		d4.2 wS/balsam fir/feather moss
		d4.3 wS/feather moss

Figure 27. continued.

8-4

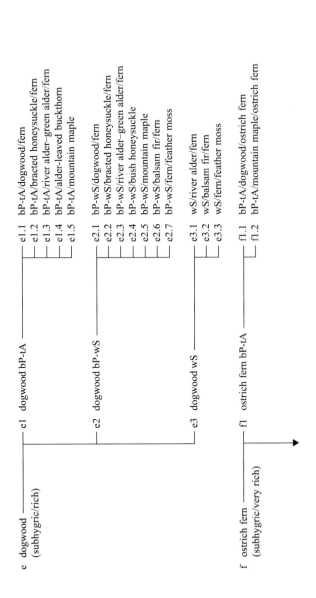

e dogwood ────── e1 dogwood bP-tA ────────── ┌─ e1.1 bP-tA/dogwood/fern
   (subhygric/rich)   ├─ e1.2 bP-tA/bracted honeysuckle/fern
                                                ├─ e1.3 bP-tA/river alder–green alder/fern
                                                ├─ e1.4 bP-tA/alder-leaved buckthorn
                                                └─ e1.5 bP-tA/mountain maple

                  e2 dogwood bP-wS ──────────── ┌─ e2.1 bP-wS/dogwood/fern
                                                 ├─ e2.2 bP-wS/bracted honeysuckle/fern
                                                 ├─ e2.3 bP-wS/river alder–green alder/fern
                                                 ├─ e2.4 bP-wS/bush honeysuckle
                                                 ├─ e2.5 bP-wS/mountain maple
                                                 ├─ e2.6 bP-wS/balsam fir/fern
                                                 └─ e2.7 bP-wS/fern/feather moss

                  e3 dogwood wS ──────────────── ┌─ e3.1 wS/river alder/fern
                                                  ├─ e3.2 wS/balsam fir/fern
                                                  └─ e3.3 wS/fern/feather moss

f ostrich fern ── f1 ostrich fern bP-tA ──────── ┌─ f1.1 bP-tA/dogwood/ostrich fern
   (subhygric/very rich)   └─ f1.2 bP-tA/mountain maple/ostrich fern

Figure 27. continued.

8-5

ECOSITE	ECOSITE PHASE	PLANT COMMUNITY TYPE
	f2 ostrich fern mM-wE-bP-gA	f2.1 mM-wE-bP-gA/pin cherry–saskatoon/ostrich fern f2.2 mM-wE-bP-gA/Manitoba maple/ostrich fern f2.3 mM-wE-bP-gA/high-bush cranberry–green ash/ostrich fern
	f3 ostrich fern bP-wS	f3.1 bP-wS/dogwood/ostrich fern
g Labrador tea–hygric (hygric/poor)	g1 Labrador tea–hygric bS-jP	g1.1 bS-jP/Labrador tea/feather moss g1.2 bS-jP/feather moss
h horsetail (hygric/rich)	h1 horsetail bP-tA	h1.1 bP-tA/horsetail
	h2 horsetail bP-wS	h2.1 bP-wS/horsetail
	h3 horsetail wS-bS	h3.1 wS-bS/horsetail h3.2 wS-bS/Labrador tea/horsetail
i gully (hygric/very rich)	i1 river alder gully	i1.1 river alder/ostrich fern

Figure 27. continued.

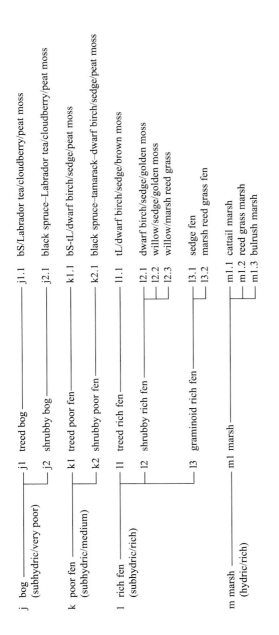

j   bog ———————————————— j1 treed bog ——————————— j1.1 bS/Labrador tea/cloudberry/peat moss
    (subhydric/very poor)   j2 shrubby bog ————————— j2.1 black spruce–Labrador tea/cloudberry/peat moss

k   poor fen ———————————— k1 treed poor fen ——————— k1.1 bS–tL/dwarf birch/sedge/peat moss
    (subhydric/medium)   k2 shrubby poor fen —————— k2.1 black spruce–tamarack–dwarf birch/sedge/peat moss

l   rich fen ————————————— l1 treed rich fen ————————— l1.1 tL/dwarf birch/sedge/brown moss
    (subhydric/rich)
                                    l2 shrubby rich fen —————— l2.1 dwarf birch/sedge/golden moss
                                                                l2.2 willow/sedge/golden moss
                                                                l2.3 willow/marsh reed grass

                                    l3 graminoid rich fen ———— l3.1 sedge fen
                                                                l3.2 marsh reed grass fen

m   marsh ———————————————— m1 marsh ————————————— m1.1 cattail marsh
    (hydric/rich)   m1.2 reed grass marsh
                                                                m1.3 bulrush marsh

Figure 27.   concluded.

8-7

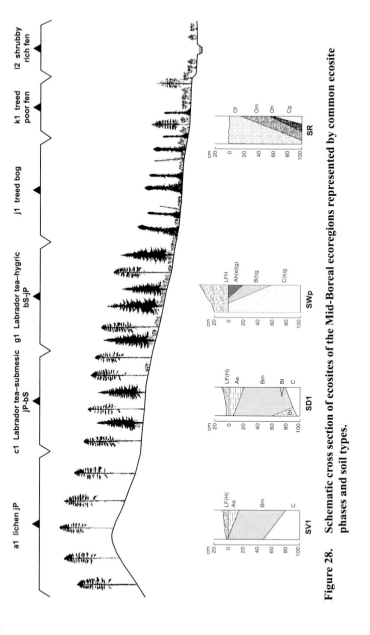

a1 lichen jP c1 Labrador tea-submesic g1 Labrador tea-hygric j1 treed bog k1 treed l2 shrubby
                jP-bS   Labrador tea-submesic   bS-jP   poor fen   rich fen

Figure 28. Schematic cross section of ecosites of the Mid-Boreal ecoregions represented by common ecosite phases and soil types.

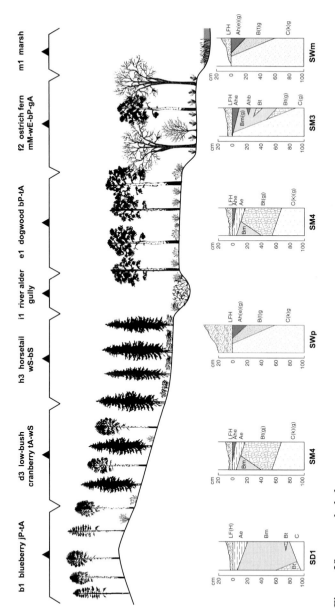

Figure 28. concluded.

GENERAL DESCRIPTION

This ecosite has dry conditions with rapidly drained acidic soils and poor nutrient status due to the coarse-textured glaciofluvial or eolian parent materials. Plants that are indicative of the nutrient-poor substrate include bearberry, reindeer lichen, bog cranberry, and blueberry. Open-canopied jack pine stands dominate this ecosite, which commonly has a carpet of lichens covering the forest floor and a thin organic layer typically less than 5 cm thick. This ecosite is more prevalent in the Mid-Boreal Upland than in the Mid-Boreal Lowland Ecoregion.

a1 lichen jP   **c1 Labrador tea–submesic jP-bS**

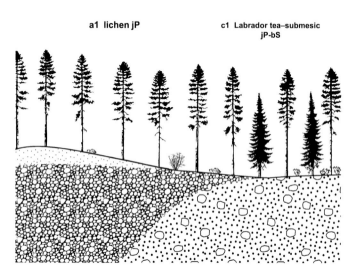

SUCCESSIONAL RELATIONSHIPS

Due to the dry nature of this ecosite, succession to black spruce is commonly slower than the fire return interval. Therefore, pine is maintained for relatively long periods and can colonize the site and dominate the canopy in a fire climax community.

MANAGEMENT INTERPRETATIONS

Drought Limitations	Excess Moisture	Rutting Hazard	Compaction Hazard	Puddling Hazard	Erosion Hazard	Frost Heave Hazard	Soil Temperature Limitations	Vegetation Competition	Windthrow Hazard	Productivity	Season of Harvest
H	L	L	L	L	L	L	L	L	L	L	W-Sp

xeric/poor

Nutrient Regime

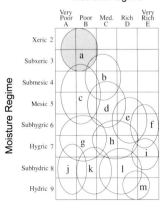

INDICATOR SPECIES

reindeer lichen
bearberry
blueberry
bog cranberry
awned hair-cap
sand heather

PRODUCTIVITY

Site Index at 50 years BH age:
jack pine: 14.9 ± 0.3 m; n = 31

Mean Annual Increment:
1.6 ± 0.2 m³/ha/yr; n = 25

SITE CHARACTERISTICS

Moisture Regime: subxeric[4], xeric[3], very xeric[2]
Nutrient Regime: poor[9]
Topographic Position: midslope[3], level[3], upper slope[2], crest[2], lower slope[1]

SOIL CHARACTERISTICS

Organic Thickness: (0–5)[9], (6–15)[1]
Humus Form: mor[8], raw moder[2]
Surface Texture: mS[6], fS[1], cS[1], LmS[1]
Effective Texture: mS[6], cS[1], fS[1], LmS[1]
Depth to Mottles/Gley: none[10]
Drainage: rapid[7], very rapid[2], well[1]
Depth to Water Table: none[10]
Parent Material: GF[4], E[2], FE[2]
Soil Subgroup: E.EB[7], O.EB[2]
Soil Association: Pn[9]
Soil Type: SV1[9]

ECOSITE PHASES (n)

a1 lichen jP (47)

---

a1 | lichen jP (n = 47)

Nutrient Regime

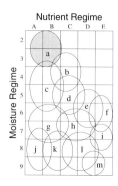

CHARACTERISTIC SPECIES

Tree
[37] jack pine

Shrub
[9] bearberry
[7] blueberry
[4] bog cranberry
[3] green alder*
[1] sand heather*

Forb
[2] wild lily-of-the-valley

Moss
[16] Schreber's moss
[1] cushion moss

Lichen
[28] reindeer lichen

SITE CHARACTERISTICS

Moisture Regime: subxeric[4], xeric[3], very xeric[2]
Nutrient Regime: poor[9]
Topographic Position: midslope[3], level[3], upper slope[2], crest[2], lower slope[1]

Stand Age: 69.9 ± 4.0 yrs; n = 34
Richness: 16.25 ± 1.05; n = 47
Diversity: 1.90 ± 0.11; n = 47

SOIL CHARACTERISTICS

Organic Thickness: (0–5)[9], (6–15)[1]
Humus Form: mor[8], raw moder[2]
Surface Texture: mS[6], cS[1], fS[1], LmS[1]
Effective Texture: mS[6], fS[1], cS[1], LmS[1]
Depth to Mottles/Gley: none[10]
Drainage: rapid[7], very rapid[2], well[1]
Depth to Water Table: none[10]
Parent Material: GF[4], E[2], FE[2]
Soil Subgroup: E.EB[7], O.EB[2]
Soil Association: Pn[9]
Soil Type: SV1[9]

PLANT COMMUNITY TYPES (n)

a1.1 jP/bearberry/lichen (23)
a1.2 jP/blueberry/lichen (14)
a1.3 jP/green alder/lichen (10)

---

* Species characteristic of the phase but occurring in <70% of the sample plots with a prominence value <20.

---

a1 lichen jP - Plant community types and approximate cover classes

		a1.1	a1.2	a1.3	
TREE	jack pine	▮	▮	▮	*Pinus banksiana*
SHRUB	green alder	▮			*Alnus crispa*
	bearberry	▮	▮	▮	*Arctostaphylos uva-ursi*
	blueberry	▮	▮	▮	*Vaccinium myrtilloides*
	bog cranberry	▮	▮	▮	*Vaccinium vitis-idaea*
	twin-flower	▪	▮		*Linnaea borealis*
	sand heather	▪	▪	▮	*Hudsonia tomentosa**
FORB	wild lily-of-the-valley		▪	▪	*Maianthemum canadense*
MOSS	Schreber's moss	▮	▮	▮	*Pleurozium schreberi*
	awned hair-cap	▪	▪	▪	*Polytrichum piliferum**
LICHEN	reindeer lichen	▮	▮	▮	*Cladina* spp.

Approximate cover classes (%): ▪ <2 ▪ 2–4.99 ▮ 5–9.99 ▬ 10–19.99 ▬ ≥20.
* Characteristic species with a prominence value <20 for all plant community types.

b | **blueberry** (n = 79)

GENERAL DESCRIPTION

This ecosite tends to be subxeric to submesic as a result of relatively coarse-textured glaciofluvial or glaciofluvial overlying morainal parent materials. This ecosite is intermediate in both moisture and nutrient regime between the lichen ecosite (a) and the low-bush cranberry ecosite (d). The blueberry ecosite has species characteristic of the lichen ecosite such as jack pine, blueberry, bearberry, and bog cranberry, and species characteristic of the low-bush cranberry ecosite, such as aspen, white spruce, wild sarsaparilla, and bunchberry. This ecosite had a higher frequency of occurrence in the Mid-Boreal Upland Ecoregion than in the Mid-Boreal Lowland Ecoregion.

d2 low-bush cranberry tA   **b1 blueberry jP-tA**

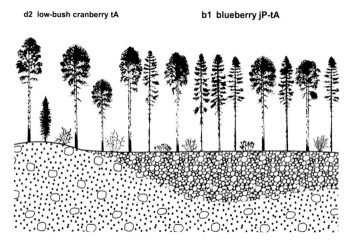

SUCCESSIONAL RELATIONSHIPS

The pine, aspen, and white birch-dominated phases of this ecosite, may in some cases, succeed to white spruce, however, the process is slow due to the dry nature of this ecosite.

MANAGEMENT INTERPRETATIONS

Drought Limitations	Excess Moisture	Rutting Hazard	Compaction Hazard	Pudding Hazard	Erosion Hazard	Frost Heave Hazard	Soil Temperature Limitations	Vegetation Competition	Windthrow Hazard	Productivity	Season of Harvest
M-H	L	L	L-M	L	L	L-M	L	M	L-M	M	A

submesic/medium

Nutrient Regime

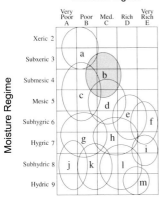

INDICATOR SPECIES

blueberry
hairy wild rye
bearberry
bog cranberry
wild sarsaparilla
Labrador tea
cream-colored vetchling

PRODUCTIVITY

Site Index at 50 years BH age:
aspen: 18.3 ± 0.4 m; n = 35
jack pine: 18.2 ± 0.4 m; n = 18
white spruce: 16.1 ± 1.0 m; n = 10
black spruce: 15.8 m; n = 1
white birch: no data

Mean Annual Increment:
2.8 ± 0.1 m³/ha/yr; n = 40

SITE CHARACTERISTICS

Moisture Regime: submesic[4], subxeric[4], mesic[1], xeric[1]
Nutrient Regime: poor[7], medium[3]
Topographic Position: midslope[3], level[3], crest[2], upper slope[2]

SOIL CHARACTERISTICS

Organic Thickness: (6–15)[5], (0–5)[5]
Humus Form: mor[9], raw moder[1]
Surface Texture: mS[4], LmS[1], SL[1], LcS[1], cS[1]
Effective Texture: mS[3], cS[1], LmS[1], C[1]
Depth to Mottles/Gley: none[9]
Drainage: rapid[4], well[4], mod. well[1]
Depth to Water Table: none[10]
Parent Material: GF[3], GF/M[1], FL[1], M[1]
Soil Subgroup: E.EB[4], BR.GL[3], O.GL[1], O.EB[1], E.DYB[1]
Soil Association: Wt[2], Kk[2], Ln[1], Pn[1], Bt[1], Bd[1]
Soil Type: SV1[4], SD1[2], SD4[2]

ECOSITE PHASES (n)

b1 blueberry jP-tA (32)
b2 blueberry tA(wB) (25)
b3 blueberry tA-wS (12)
b4 blueberry wS(jP) (10)

ECOSITE

Nutrient Regime

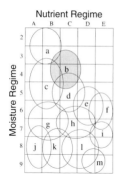

CHARACTERISTIC SPECIES

Tree
[29] jack pine
[26] aspen
[4] white spruce
[2] white birch[x]

Shrub
[11] green alder
[6] blueberry
[4] bog cranberry
[4] bearberry
[4] twin-flower
[3] prickly rose
[3] Labrador tea
[2] aspen*

Forb
[4] bunchberry
[3] wild sarsaparilla
[2] stiff club-moss*
[2] wild lily-of-the-valley
[1] northern starflower
[1] cream-colored vetchling*

Grass
[2] hairy wild rye*

Moss
[21] Schreber's moss
[8] stair-step moss
[3] knight's plume moss

Lichen
[3] reindeer lichen*

SITE CHARACTERISTICS

Moisture Regime: submesic[4], subxeric[3], mesic[2], xeric[1]
Nutrient Regime: poor[6], medium[4]
Topographic Position: level[3], midslope[2], crest[2], upper slope[2]

Stand Age: 85.7 ± 5.1 yrs; n = 19
Richness: 24.81 ± 1.24; n = 32
Diversity: 2.84 ± 0.12; n = 32

SOIL CHARACTERISTICS

Organic Thickness: (6–15)[6], (0–5)[4]
Humus Form: mor[9]
Surface Texture: mS[3], SL[3], LmS[2], LcS[1]
Effective Texture: mS[3], C[2], SCL[1], SC[1]
Depth to Mottles/Gley: none[9]
Drainage: well[4], rapid[4], mod. well[1]
Depth to Water Table: none[10]
Parent Material: M[2], GF/M[2], GF[2], FL[1]
Soil Subgroup: E.EB[3], BR.GL[3], O.GL[1], E.DYB[1]
Soil Association: Ln[3], Wt[2], Bt[2], Kk[1]
Soil Type: SV1[3], SD4[2], SM4[2], SD1[1]

PLANT COMMUNITY TYPES (n)

b1.1 jP-tA/blueberry–bearberry (12)
b1.2 jP-tA/blueberry–green alder (20)

---

* Species characteristic of the phase but occurring in <70% of the sample plots with a prominence value <20.
[x] Tree may be dominant in some plots.

b1 blueberry jP-tA - Plant community types and approximate cover classes

		b1.1	b1.2	
TREE	jack pine	▮	▮	*Pinus banksiana*
	aspen	▮	▮	*Populus tremuloides*
	white spruce	▪	▮	*Picea glauca*
	white birch	▪	▪	*Betula papyrifera* *
SHRUB	green alder		▮	*Alnus crispa*
	blueberry	▮	▮	*Vaccinium myrtilloides*
	twin-flower	▮	▪	*Linnaea borealis*
	bog cranberry	▪	▪	*Vaccinium vitis-idaea*
	Labrador tea		▪	*Ledum groenlandicum*
	bearberry	▪	▪	*Arctostaphylos uva-ursi*
	white spruce	▪	▪	*Picea glauca*
	prickly rose	▪	▪	*Rosa acicularis*
	white birch	▪	▪	*Betula papyrifera* *
FORB	bunchberry	▮	▪	*Cornus canadensis*
	wild sarsaparilla	▪		*Aralia nudicaulis*
	cream-colored vetchling		▪	*Lathyrus ochroleucus* *
GRASS	hairy wild rye	▪	▪	*Elymus innovatus* *
MOSS	Schreber's moss	▮	▮	*Pleurozium schreberi*
	stair-step moss	▮	▪	*Hylocomium splendens*
	knight's plume moss	▪		*Ptilium crista-castrensis*
LICHEN	reindeer lichen	▪		*Cladina* spp.

Approximate cover classes (%): ▪ <2   ▪ 2–4.99   ■ 5–9.99   ▬ 10–19.99   ▬ ≥20.
* Characteristic species with a prominence value <20 for all plant community types.

b2 | blueberry tA(wB) (n = 25)

Nutrient Regime

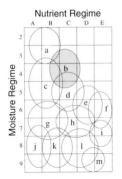

CHARACTERISTIC SPECIES

Tree
[51] aspen
[12] white birch
[1] white spruce*

Shrub
[12] blueberry
[10] green alder
[7] bearberry
[7] Labrador tea
[4] prickly rose
[4] twin-flower
[3] bog cranberry
[3] saskatoon*

Forb
[8] bunchberry
[3] wild sarsaparilla
[2] wild lily-of-the-valley

Grass
[2] hairy wild rye*

SITE CHARACTERISTICS

Moisture Regime: submesic[5], subxeric[4]
Nutrient Regime: poor[6], medium[4]
Topographic Position: midslope[3], level[3], upper slope[2], lower slope[1]

Stand Age: 62.7 ± 2.8 yrs; n = 14
Richness: 25.80 ± 1.25; n = 25
Diversity: 2.94 ± 0.11; n = 25

SOIL CHARACTERISTICS

Organic Thickness: (6–15)[5], (0–5)[5]
Humus Form: mor[8], raw moder[2]
Surface Texture: mS[4], LfS[2], cS[1], Si[1], LcS[1]
Effective Texture: mS[4], LfS[2], LmS[1]
Depth to Mottles/Gley: none[10]
Drainage: rapid[5], well[4], mod. well[1]
Depth to Water Table: none[10]
Parent Material: GF[3], FL[2], GF/M[1]
Soil Subgroup: E.EB[3], BR.GL[3], O.GL[2], O.EB[1], E.DYB[1]
Soil Association: Wt[4], Pn[1], Ln[1], Kk[1], Hw[1], Bx[1], Bd[1]
Soil Type: SV1[4], SD1[4], SD4[2]

PLANT COMMUNITY TYPES (n)

b2.1 tA(wB)/blueberry–bearberry (8)
b2.2 tA(wB)/blueberry–green alder (12)
b2.3 tA(wB)/blueberry–Labrador tea (5)

---

* Species characteristic of the phase but occurring in <70% of the sample plots with a prominence value <20.

b2 blueberry tA(wB) - Plant community types and approximate cover classes

	common name	scientific name	b2.1	b2.2	b2.3
TREE	aspen	*Populus tremuloides*			
	white birch	*Betula papyrifera*			
	white spruce	*Picea glauca**			
SHRUB	Labrador tea	*Ledum groenlandicum*			
	green alder	*Alnus crispa*			
	blueberry	*Vaccinium myrtilloides*			
	bearberry	*Arctostaphylos uva-ursi*			
	prickly rose	*Rosa acicularis*			
	twin-flower	*Linnaea borealis*			
	saskatoon	*Amelanchier alnifolia*			
	pin and choke cherry	*Prunus* spp.			
	bog cranberry	*Vaccinium vitis-idaea*			
	aspen	*Populus tremuloides*			
	white spruce	*Picea glauca**			
FORB	bunchberry	*Cornus canadensis*			
	spreading dogbane	*Apocynum androsaemifolium**			
	wild sarsaparilla	*Aralia nudicaulis*			
	wild lily-of-the-valley	*Maianthemum canadense*			
GRASS	marsh reed grass	*Calamagrostis canadensis*			
	hairy wild rye	*Elymus innovatus*			
MOSS	Schreber's moss	*Pleurozium schreberi*			

Approximate cover classes (%): ■ <2   ■ 2–4.99   ■ 5–9.99   ■ 10–19.99   ■ ≥20.
* Characteristic species with a prominence value <20 for all plant community types.

8-19

b3 | blueberry tA-wS (n = 12)

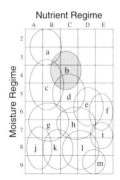

Nutrient Regime

Moisture Regime

CHARACTERISTIC SPECIES

Tree
[28] aspen
[24] white spruce

Shrub
[17] blueberry
[14] bearberry
[7] bog cranberry
[5] prickly rose
[4] white spruce
[4] Labrador tea
[3] green alder
[2] aspen
[2] twin-flower
[2] saskatoon*

Forb
[5] bunchberry
[2] wild sarsaparilla
[2] wild lily-of-the-valley
[1] cream-colored vetchling*

Moss
[14] Schreber's moss
[5] stair-step moss

Lichen
[2] reindeer lichen*

SITE CHARACTERISTICS

Moisture Regime: submesic[5], subxeric[4], xeric[1]
Nutrient Regime: poor[8], medium[2]
Topographic Position: midslope[5], upper slope[2], crest[2], level[1], depression[1]

Stand Age: 88.8 ± 8.3 yrs; n = 5
Richness: 26.42 ± 1.42; n = 12
Diversity: 3.11 ± 0.13; n = 12

SOIL CHARACTERISTICS

Organic Thickness: (0–5)[6], (6–15)[4]
Humus Form: mor[7], raw moder[3]
Surface Texture: mS[4], LmS[3], fS[1], SiS[1], LfS[1], LcS[1]
Effective Texture: LmS[3], mS[2], fS[1], cS[1], SiS[1], Si[1], SL[1], LfS[1]
Depth to Mottles/Gley: none[8], >100[1], (51–100)[1]
Drainage: well[5], rapid[5]
Depth to Water Table: none[10]
Parent Material: GF[5], F[2], GF/M[1], FE[1], E/GF[1]
Soil Subgroup: E.EB[4], BR.GL[3], E.DYB[2], O.GL[1], GLE.EB[1]
Soil Association: Pn[3], Bd[3], Bt[2], Av[2]
Soil Type: SV1[4], SD1[4], SD2[2]

PLANT COMMUNITY TYPES (n)

b3.1 tA-wS/blueberry–bearberry (7)
b3.2 tA-wS/blueberry–green alder (3)
b3.3 tA-wS/blueberry–Labrador tea (2)

---

* Species characteristic of the phase but occurring in <70% of the sample plots with a prominence value <20.

ECOSITE PHASE

b3 blueberry tA-wS - Plant community types and approximate cover classes

		b3.1	b3.2	b3.3	
TREE	aspen				Populus tremuloides
	white spruce				Picea glauca
SHRUB	blueberry				Vaccinium myrtilloides
	bearberry				Arctostaphylos uva-ursi
	Labrador tea				Ledum groenlandicum
	green alder				Alnus crispa
	bog cranberry				Vaccinium vitis-idaea
	white spruce				Picea glauca
	prickly rose				Rosa acicularis
	aspen				Populus tremuloides
	twin-flower				Linnaea borealis
	low-bush cranberry				Viburnum edule
	pin and choke cherry				Prunus spp.
	saskatoon				Amelanchier alnifolia*
FORB	bunchberry				Cornus canadensis
	wild sarsaparilla				Aralia nudicaulis
	fireweed				Epilobium angustifolium
	wild lily-of-the-valley				Maianthemum canadense
	cream-colored vetchling				Lathyrus ochroleucus*
GRASS	marsh reed grass				Calamagrostis canadensis
	hairy wild rye				Elymus innovatus*
MOSS	Schreber's moss				Pleurozium schreberi
	stair-step moss				Hylocomium splendens
LICHEN	reindeer lichen				Cladina spp.*

Approximate cover classes (%): ■ <2 ■ 2-4.99 ■ 5-9.99 ▬ 10-19.99 ▬▬ ≥20.

* Characteristic species with a prominence value <20 for all plant community types.

b4 blueberry wS(jP) (n = 10)

Nutrient Regime

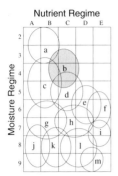

CHARACTERISTIC SPECIES

Tree
[33] white spruce
[17] jack pine

Shrub
[12] bearberry
[11] green alder
[7] blueberry
[3] white spruce
[3] bog cranberry
[2] Canada buffalo-berry*
[2] prickly rose
[1] twin-flower
[1] saskatoon*

Forb
[2] wild lily-of-the-valley
[1] bastard toad-flax

Moss
[25] Schreber's moss
[6] stair-step moss
[6] knight's plume moss

Lichen
[7] reindeer lichen

SITE CHARACTERISTICS

Moisture Regime: subxeric[7], xeric[3]
Nutrient Regime: poor[10]
Topographic Position: crest[5], midslope[3], level[2]

Stand Age: 108.6 ± 8.6 yrs; n = 4
Richness: 19.90 ± 1.75; n = 10
Diversity: 2.20 ± 0.22; n = 10

SOIL CHARACTERISTICS

Organic Thickness: (0–5)[7], (6–15)[3]
Humus Form: mor[10]
Surface Texture: mS[3], cS[3], SiL[3]
Effective Texture: cS[5], mS[2], SiL[2], Si[2]
Depth to Mottles/Gley: none[10]
Drainage: well[3], rapid[3], mod. well[2], very rapid[1]
Depth to Water Table: none[10]
Parent Material: GF[5]
Soil Subgroup: E.EB[6], O.EB[3], O.GL[1]
Soil Association: Kk[10]
Soil Type: SV1[7]

PLANT COMMUNITY TYPES (n)

b4.1 wS(jP)/blueberry–bearberry (7)
b4.2 wS(jP)/blueberry–green alder (3)

---

* Species characteristic of the phase but occurring in <70% of the sample plots with a prominence value <20.

ECOSITE PHASE

b4 blueberry wS(jP) - Plant community types and approximate cover classes

		b4.1	b4.2	
TREE	white spruce	▮	▮	*Picea glauca*
	jack pine	▮	▮	*Pinus banksiana*
SHRUB	green alder		▮	*Alnus crispa*
	bearberry	▮	▪	*Arctostaphylos uva-ursi*
	blueberry	▮	▮	*Vaccinium myrtilloides*
	white spruce	▪	▪	*Picea glauca*
	bog cranberry	▪		*Vaccinium vitis-idaea*
	Canada buffalo-berry	▪		*Shepherdia canadensis*
	prickly rose	▪		*Rosa acicularis* *
	saskatoon		▪	*Amelanchier alnifolia* *
MOSS	Schreber's moss	▮	▮	*Pleurozium schreberi*
	knight's plume moss		▮	*Ptilium crista-castrensis*
	stair-step moss	▪	▪	*Hylocomium splendens*
LICHEN	reindeer lichen	▪		*Cladina* spp.

Approximate cover classes (%): ▪ <2   ▪ 2–4.99   ▪ 5–9.99   ▮ 10–19.99   ▮ ≥20.
* Characteristic species with a prominence value <20 for all plant community types.

GENERAL DESCRIPTION

This ecosite has a subxeric to subhygric nutrient-poor substrate. Labrador tea and bog cranberry are indicative of the relatively acidic surface soil conditions. It occurs in upland (midslope and upper slope) or level topographic positions dominantly on morainal or glaciofluvial parent materials. There is commonly a two-tiered even-aged canopy where the faster-growing jack pine comprise the higher layer and the slower-growing black spruce form a secondary canopy below the pine. While the Labrador tea–submesic ecosite has plant community types similar to the Labrador tea–hygric ecosite (g), the submesic ecosite tends to occur in upper topographic positions, has no mottles within the top 25 cm of soil, and a thinner organic layer. This ecosite covers a higher proportion of the landscape area in the Mid-Boreal Upland Ecoregion than in the Mid-Boreal Lowland Ecoregion.

c1 Labrador tea–submesic jP-bS   **g1 Labrador tea–hygric bS-jP**

SUCCESSIONAL RELATIONSHIPS

Successionally mature stands that develop on these ecosites may be dominated by black spruce. Residual pine occurring in the climax community are generally very old. The successionally mature stage is rare due to a high frequency of fire.

MANAGEMENT INTERPRETATIONS

Drought Limitations	Excess Moisture	Rutting Hazard	Compaction Hazard	Puddling Hazard	Erosion Hazard	Frost Heave Hazard	Soil Temperature Limitations	Vegetation Competition	Windthrow Hazard	Productivity	Season of Harvest
L-H	L	L-M	M	L-M	M	L-M	L	L	L	L-M	A

submesic/poor

Nutrient Regime

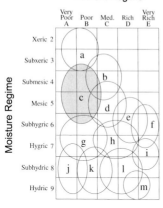

INDICATOR SPECIES

jack pine
black spruce
Labrador tea
bog cranberry
Schreber's moss

PRODUCTIVITY

Site Index at 50 years BH age:
jack pine: 16.1 ± 0.4 m; n = 25
black spruce: 12.2 ± 1.3 m; n = 4
white birch: 9.2 m; n = 1

Mean Annual Increment:
2.7 ± 0.2 m³/ha/yr; n = 23

SITE CHARACTERISTICS

Moisture Regime: submesic[6], mesic[3], subxeric[1]
Nutrient Regime: poor[8], medium[2]
Topographic Position: midslope[4], level[2], upper slope[2], crest[1]

SOIL CHARACTERISTICS

Organic Thickness: (6–15)[6], (0–5)[4]
Humus Form: mor[9], raw moder[1]
Surface Texture: mS[2], SiS[2], SL[2], LmS[2], SiL[1]
Effective Texture: mS[3], SC[2], LmS[2], SL[1], SiS[1], CL[1]
Depth to Mottles/Gley: none[10]
Drainage: well[5], rapid[3], mod. well[2]
Depth to Water Table: none[10]
Parent Material: M[4], GF[2], GF/M[1], FL[1]
Soil Subgroup: BR.GL[3], O.GL[3], E.EB[2]
Soil Association: Ln[5], Bt[2], Wt[1]
Soil Type: SD1[3], SD4[2], SM4[1], SD2[1], SV1[1]

ECOSITE PHASES (n)

c1 Labrador tea–submesic jP-bS (44)

c1 Labrador tea–submesic jP-bS (n = 44)

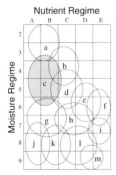

Nutrient Regime

Moisture Regime

CHARACTERISTIC SPECIES

Tree
[44] jack pine
[11] black spruce

Shrub
[11] green alder
[6] Labrador tea
[4] bog cranberry
[4] blueberry
[3] twin-flower
[3] black spruce*

Forb
[2] bunchberry
[1] wild lily-of-the-valley
[1] northern starflower*

Moss
[58] Schreber's moss
[8] stair-step moss
[4] knight's plume moss
[1] cushion moss

Lichen
[2] reindeer lichen*

SITE CHARACTERISTICS

Moisture Regime: submesic[6], mesic[3], subxeric[1]
Nutrient Regime: poor[8], medium[2]
Topographic Position: midslope[4], level[2], upper slope[2], crest[1]

Stand Age: 75.9 ± 4.5 yrs; n = 25
Richness: 17.36 ± 0.96; n = 44
Diversity: 1.78 ± 0.07; n = 44

SOIL CHARACTERISTICS

Organic Thickness: (6–15)[6], (0–5)[4]
Humus Form: mor[9], raw moder[1]
Surface Texture: mS[2], SiS[2], SL[2], LmS[2], SiL[1]
Effective Texture: mS[3], SC[2], LmS[2], SL[1], SiS[1], CL[1]
Depth to Mottles/Gley: none[10]
Drainage: well[5], rapid[3], mod. well[2]
Depth to Water Table: none[10]
Parent Material: M[4], GF[2], GF/M[1], FL[1]
Soil Subgroup: BR.GL[3], O.GL[3], E.EB[2]
Soil Association: Ln[5], Bt[2], Wt[1]
Soil Type: SD1[3], SD4[2], SM4[1], SD2[1], SV1[1]

PLANT COMMUNITY TYPES (n)

c1.1 jP-bS/Labrador tea/feather moss (8)
c1.2 jP-bS/green alder/feather moss (24)
c1.3 jP-bS/feather moss (12)

---

* Species characteristic of the phase but occurring in <70% of the sample plots with a prominence value <20.

c1 Labrador tea–submesic jP-bS - Plant community types and approximate cover classes

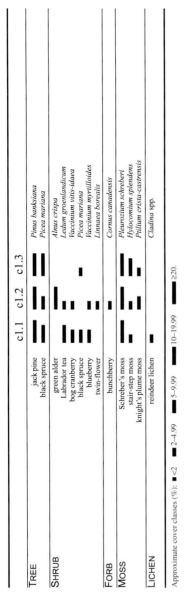

		c1.1	c1.2	c1.3	
TREE	jack pine				*Pinus banksiana*
	black spruce				*Picea mariana*
SHRUB	green alder				*Alnus crispa*
	Labrador tea				*Ledum groenlandicum*
	bog cranberry				*Vaccinium vitis-idaea*
	black spruce				*Picea mariana*
	blueberry				*Vaccinium myrtilloides*
	twin-flower				*Linnaea borealis*
FORB	bunchberry				*Cornus canadensis*
MOSS	Schreber's moss				*Pleurozium schreberi*
	stair-step moss				*Hylocomium splendens*
	knight's plume moss				*Ptilium crista-castrensis*
LICHEN	reindeer lichen				*Cladina* spp.

Approximate cover classes (%): ▪ <2   ■ 2–4.99   ■ 5–9.99   ■ 10–19.99   ■ ≥20.

GENERAL DESCRIPTION

This is the reference ecosite for the Mid-Boreal ecoregions because it has a mesic moisture regime and a medium nutrient regime. Generally, these ecosites have moderately fine to fine-textured till or glaciolacustrine parent materials. This ecosite covers a relatively large portion of the Mid-Boreal Upland Ecoregion. It is far less common in the Mid-Boreal Lowland Ecoregion because a high proportion of the landscape of the Lowland Ecoregion consists of low-lying wetland areas. Mountain maple (*Acer spicatum*) and bush honeysuckle (*Diervilla lonicera*) community types of the low-bush cranberry ecosite are only in the eastern portion of Saskatchewan.

d3 low-bush cranberry tA-wS   **h3 horsetail wS-bS**

SUCCESSIONAL RELATIONSHIPS

Pioneer deciduous tree species such as aspen, balsam poplar, and white birch are replaced by white spruce and balsam fir as these sites develop successionally. Along with a change in canopy composition is a change in understory structure and understory species composition and abundance. Generally, as a stand successionally matures, the coniferous canopy cover increases, and understory species structure and diversity declines. This results in stands with low cover of shrub, forb, and grass species and high moss cover.

MANAGEMENT INTERPRETATIONS

Drought Limitations	Excess Moisture	Rutting Hazard	Compaction Hazard	Puddling Hazard	Erosion Hazard	Frost Heave Hazard	Soil Temperature Limitations	Vegetation Competition	Windthrow Hazard	Productivity	Season of Harvest
L-M	L	M	M	M	M-H	M	L	H	L	M-H	A

mesic/medium

Nutrient Regime

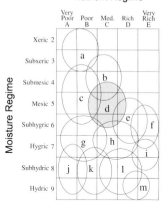

INDICATOR SPECIES

low-bush cranberry
dewberry
wild sarsaparilla

PRODUCTIVITY

Site Index at 50 years BH age:
aspen: 20.0 ± 0.2 m; n = 98
white spruce: 19.7 ± 0.4 m; n = 79
jack pine: 18.6 ± 0.9 m; n = 7
white birch: 16.8 ± 0.8 m; n = 7
balsam poplar: 18.9 ± 0.6 m; n = 5
black spruce: 14.9 ± 0.6 m; n = 5
balsam fir: 18.5 ± 1.4 m; n = 4

Mean Annual Increment:
3.8 ± 0.1 m³/ha/yr; n = 144

SITE CHARACTERISTICS

Moisture Regime: mesic⁶, submesic², subhygric¹
Nutrient Regime: medium⁷, poor¹, rich¹
Topographic Position: midslope⁴, upper slope², level², lower slope¹

SOIL CHARACTERISTICS

Organic Thickness: (6–15)⁸, (0–5)²
Humus Form: mor⁷, raw moder³
Surface Texture: SiL², SL², Si¹, LmS¹
Effective Texture: C³, SiC¹, SCL¹
Depth to Mottles/Gley: none⁸, (0–25)¹
Drainage: mod. well⁵, well⁴
Depth to Water Table: none¹⁰
Parent Material: M⁴, GL¹, GF¹
Soil Subgroup: O.GL⁴, BR.GL², E.EB¹
Soil Association: Bt², Ln², Do¹, Wv¹
Soil Type: SM4⁶, SD4¹

ECOSITE PHASES (n)

d1 low-bush cranberry jP-bS-tA (11)
d2 low-bush cranberry tA (123)
d3 low-bush cranberry tA-wS (136)
d4 low-bush cranberry wS (31)

Nutrient Regime

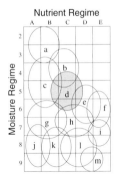

CHARACTERISTIC SPECIES

Tree
[28] jack pine
[4] aspen[x]
[3] white birch*
[3] black spruce[x]

Shrub
[14] green alder
[5] bush honeysuckle
[2] prickly rose
[1] low-bush cranberry

Forb
[15] stiff club-moss
[9] wild sarsaparilla
[5] bunchberry
[5] dewberry
[2] wild strawberry
[1] palmate-leaved coltsfoot
[1] northern starflower
[1] wild lily-of-the-valley
[+] shield fern*
[+] lady fern*

Grass
[3] hairy wild rye

Moss
[31] Schreber's moss
[8] stair-step moss
[7] knight's plume moss

SITE CHARACTERISTICS

Moisture Regime: mesic[6], submesic[2], subhygric[2]
Nutrient Regime: medium[8], rich[1], poor[1]
Topographic Position: midslope[5] level[3], upper slope[2], crest[1]

Stand Age: 77.8 ± 12.0 yrs; n = 4
Richness: 30.55 ± 1.94; n = 11
Diversity: 2.83 ± 0.20; n = 11

SOIL CHARACTERISTICS

Organic Thickness: $(6–15)^9$, $(0–5)^1$
Humus Form: mor[10]
Surface Texture: SL[5], SiCL[3], Si[3]
Effective Texture: SiC[3], SL[3], SC[3], C[3]
Depth to Mottles/Gley: none[5], $(51–100)^3$, $(0–25)^3$
Drainage: well[5], mod. well[4], imperfect[2]
Depth to Water Table: none[10]
Parent Material: M[7], GF/M[2], GF[1]
Soil Subgroup: O.GL[5], BR.GL[2], O.LG[1], GLE.EB[1], E.EB[1]
Soil Association: Ln[8], Bt[3]
Soil Type: SM4[8], SD2[3]

PLANT COMMUNITY TYPES (n)

d1.1 jP-bS-tA/stiff club-moss–fern (11)

---

* Species characteristic of the phase but occurring in <70% of the sample plots with a prominence value <20.
[x] Tree may be dominant in some plots.

d1 low-bush cranberry jP-bS-tA - Plant community types and approximate cover classes

		d1.1	
TREE			
	jack pine	▮	Pinus banksiana
	aspen		Populus tremuloides
	white birch	▮	Betula papyrifera*
	black spruce	▮	Picea mariana*
SHRUB			
	green alder	▮	Alnus crispa
	bush honeysuckle	▮	Diervilla lonicera
	prickly rose	▪	Rosa acicularis*
	low-bush cranberry	▪	Viburnum edule*
FORB			
	stiff club-moss	▮	Lycopodium annotinum
	wild sarsaparilla	▮	Aralia nudicaulis
	bunchberry	▮	Cornus canadensis
	dewberry	▮	Rubus pubescens
	wild strawberry	▪	Fragaria virginiana*
	palmate-leaved coltsfoot	▪	Petasites palmatus*
	northern starflower	▪	Trientalis borealis*
	wild lily-of-the-valley	▪	Maianthemum canadense*
	shield fern	▪	Dryopteris carthusiana*
	lady fern	▪	Athyrium filix-femina*
GRASS			
	hairy wild rye	▮	Elymus innovatus
MOSS			
	Schreber's moss	▮	Pleurozium schreberi
	stair-step moss	▮	Hylocomium splendens
	knight's plume moss	▮	Ptilium crista-castrensis

Approximate cover classes (%): ▪ <2   ▮ 2–4.99   ▬ 5–9.99   ▬ 10–19.99   ▬ ≥20.

* Characteristic species with a prominence value <20 for all plant community types.

d2 low-bush cranberry tA (n = 123)

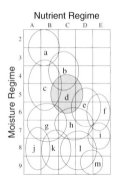

Nutrient Regime / Moisture Regime

SITE CHARACTERISTICS

Moisture Regime: mesic[6], submesic[3], subhygric[1]
Nutrient Regime: medium[7], poor[1], rich[1]
Topographic Position: midslope[4], upper slope[3], level[2], lower slope[1]

Stand Age: 63.4 ± 2.7 yrs; n = 60
Richness: 31.85 ± 0.59; n = 123
Diversity: 3.25 ± 0.05; n = 123

SOIL CHARACTERISTICS

Organic Thickness: (6–15)[8], (0–5)[2]
Humus Form: mor[6], raw moder[4]
Surface Texture: SL[2], SiL[2], Si[1], LmS[1]
Effective Texture: C[2], SiC[2], SCL[1]
Depth to Mottles/Gley: none[9], (0–25)[1]
Drainage: mod. well[5], well[4]
Depth to Water Table: none[10]
Parent Material: M[4], GF[1], GL[1], GF/M[1]
Soil Subgroup: O.GL[4], BR.GL[2], E.EB[1]
Soil Association: Bt[3], Wv[1], Ln[1], Pn[1]
Soil Type: SM4[6], SD4[2], SD1[1]

CHARACTERISTIC SPECIES

Tree
[53] aspen
[4] white birch[x]
[2] balsam poplar[x]
[1] white spruce*

Shrub
[12] beaked hazelnut
[9] green alder
[8] prickly rose
[6] low-bush cranberry
[4] twin-flower
[3] pin and choke cherry*
[2] willow*

Forb
[11] wild sarsaparilla
[5] bunchberry
[4] dewberry
[3] fireweed
[2] cream-colored vetchling
[2] wild strawberry
[1] palmate-leaved coltsfoot
[1] wild lily-of-the-valley
[1] common pink wintergreen
[1] Lindley's aster

Grass
[3] marsh reed grass

PLANT COMMUNITY TYPES (n)

d2.1 tA/pin cherry–saskatoon (9)
d2.2 tA/beaked hazelnut (31)
d2.3 tA/green alder (39)
d2.4 tA/low-bush cranberry–prickly rose (26)
d2.5 tA/willow (3)
d2.6 tA/bush honeysuckle (2)
d2.7 tA/mountain maple (2)
d2.8 tA/forb (11)

---

* Species characteristic of the phase but occurring in <70% of the sample plots with a prominence value <20.
[x] Tree may be dominant in some plots.

d2 low-bush cranberry tA

Plant community types on following pages

d2 low-bush cranberry tA - Plant community types and approximate cover classes

		d2.1	d2.2	d2.3	d2.4	d2.5	d2.6	d2.7	d2.8
TREE									
aspen	Populus tremuloides	▮	▮	▮	▮	▪	▮	▮	
white birch	Betula papyrifera		▪					▮	
balsam poplar	Populus balsamifera*	▪	▪					▪	
white spruce	Picea glauca								▮
jack pine	Pinus banksiana								▪
SHRUB									
mountain maple	Acer spicatum								
beaked hazelnut	Corylus cornuta	▪	▮	▮				▮	
willow	Salix spp.	▮					▪		
green alder	Alnus spp.	▮	▪	▮	▮	▮	▪		
pin and choke cherry	Prunus spp.		▪						
bush honeysuckle	Diervilla lonicera	▪		▪			▮		
low-bush cranberry	Viburnum edule	▮	▮	▮	▮	▮			
prickly rose	Rosa acicularis	▮	▮						▪
saskatoon	Amelanchier alnifolia								
blueberry	Vaccinium myrtilloides								
white birch	Betula papyrifera						▮		
dogwood	Cornus stolonifera						▮		
twin-flower	Linnaea borealis	▪	▪	▪	▪			▪	
river alder	Alnus tenuifolia					▪			▪
aspen	Populus tremuloides					▪			
Canada buffalo-berry	Shepherdia canadensis								▮
balsam poplar	Populus balsamifera*	▪	▪	▪	▪	▪		▪	▪
FORB									
wild sarsaparilla	Aralia nudicaulis	▮	▮	▮	▮	▮	▮	▮	▮
dewberry	Rubus pubescens	▮	▮		▮	▮	▮	▪	▮
fireweed	Epilobium angustifolium	▪	▮	▪	▮	▪			▮
spreading dogbane	Apocynum androsaemifolium	▮	▪	▪	▪		▪		▮
bunchberry	Cornus canadensis								▮
wild strawberry	Fragaria virginiana								▮
palmate-leaved coltsfoot	Petasites palmatus		▪						▮
cream-colored vetchling	Lathyrus ochroleucus								▮

d2 low-bush cranberry tA - Plant community types and approximate cover classes (concluded)

	d2.1	d2.2	d2.3	d2.4	d2.5	d2.6	d2.7	d2.8	
GRASS									
marsh reed grass	■		■	■					*Calamagrostis canadensis*
hairy wild rye	■	■							*Elymus innovatus*
MOSS									
Schreber's moss								■	*Pleurozium schreberi*

Approximate cover classes (%): ■ <2   ■ 2–4.99   ■ 5–9.99   ■ 10–19.99   ■ ≥20.
* Characteristic species with a prominence value <20 for all plant community types.

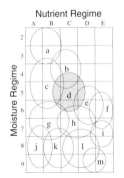

Nutrient Regime

Moisture Regime

SITE CHARACTERISTICS

Moisture Regime: mesic[8], subhygric[1], submesic[1]
Nutrient Regime: medium[8], poor[1], rich[1]
Topographic Position: midslope[4], upper slope[2], level[2], lower slope[1]

Stand Age: 81.8 ± 2.6 yrs; n = 76
Richness: 29.30 ± 0.65; n = 136
Diversity: 3.00 ± 0.06; n = 136

SOIL CHARACTERISTICS

Organic Thickness: (6–15)[8], (0–5)[2]
Humus Form: mor[7], raw moder[3]
Surface Texture: SiL[3], SL[1], Si[1], LmS[1], L[1]
Effective Texture: C[3], SiC[1], SiCL[1], SCL[1], SC[1], LmS[1]
Depth to Mottles/Gley: none[8], (0–25)[1]
Drainage: mod. well[5], well[3]
Depth to Water Table: none[10]
Parent Material: M[4], GL[1], GL/M[1]
Soil Subgroup: O.GL[4], BR.GL[2], E.EB[1]
Soil Association: Ln[3], Bt[2], Do[1], Wv[1]
Soil Type: SM4[7], SM1[1]

CHARACTERISTIC SPECIES

Tree
[29] aspen
[28] white spruce
[4] white birch[x]
[3] balsam fir[x]
[3] black spruce[x]
[2] balsam poplar[x]

Shrub
[6] low-bush cranberry
[4] green alder
[4] balsam fir
[3] prickly rose
[3] twin-flower
[1] beaked hazelnut*
[1] bush honeysuckle*
[1] mountain maple*
[+] pin and choke cherry*
[+] saskatoon*

Forb
[7] wild sarsaparilla
[4] bunchberry
[3] dewberry
[2] bishop's-cap
[2] tall lungwort
[1] palmate-leaved coltsfoot
[1] wild lily-of-the-valley
[1] common pink wintergreen

Moss
[11] stair-step moss
[7] Schreber's moss

PLANT COMMUNITY TYPES (n)

d3.1 tA-wS/pin cherry–saskatoon (3)
d3.2 tA-wS/beaked hazelnut (3)
d3.3 tA-wS/green alder (19)
d3.4 tA-wS/low-bush cranberry–prickly rose (38)
d3.5 tA-wS/bush honeysuckle (2)
d3.6 tA-wS/mountain maple (1)
d3.7 tA-wS/forb (23)
d3.8 tA-wS/balsam fir/ feather moss (19)
d3.9 tA-wS/feather moss (28)

* Species characteristic of the phase but occurring in <70% of the sample plots with a prominence value <20.
[x] Tree may be dominant in some plots.

d3 low-bush cranberry tA-wS

Plant community types on following pages

d3 low-bush cranberry tA-wS - Plant community types and approximate cover classes

		d3.1	d3.2	d3.3	d3.4	d3.5	d3.6	d3.7	d3.8	d3.9	
TREE	white spruce										*Picea glauca*
	aspen										*Populus tremuloides*
	white birch										*Betula papyrifera*
	balsam fir										*Abies balsamea*
	black spruce										*Picea mariana*
	balsam poplar										*Populus balsamifera*
SHRUB	mountain maple										*Acer spicatum*
	bush honeysuckle										*Diervilla lonicera*
	beaked hazelnut										*Corylus cornuta*
	balsam fir										*Abies balsamea*
	green alder										*Alnus crispa*
	low-bush cranberry										*Viburnum edule*
	pin and choke cherry										*Prunus* spp.
	Canada buffalo-berry										*Shepherdia canadensis*
	prickly rose										*Rosa acicularis*
	twin-flower										*Linnaea borealis*
	white birch										*Betula papyrifera*
	saskatoon										*Amelanchier alnifolia*
	white spruce										*Picea glauca*
FORB	wild sarsaparilla										*Aralia nudicaulis*
	bishop's-cap										*Mitella nuda*
	bunchberry										*Cornus canadensis*
	stiff club-moss										*Lycopodium annotinum*
	wild strawberry										*Fragaria virginiana*
	dewberry										*Rubus pubescens*
	tall lungwort										*Mertensia paniculata*
	kidney-leaved violet										*Viola renifolia*
GRASS	marsh reed grass										*Calamagrostis canadensis*

d3 low-bush cranberry tA-wS - Plant community types and approximate cover classes (concluded)

	d3.1	d3.2	d3.3	d3.4	d3.5	d3.6	d3.7	d3.8	d3.9	
MOSS										
stair-step moss	▮	▮	▮	▮	▮			▮	▮	*Hylocomium splendens*
Schreber's moss		▮	▮	▮				▮	▮	*Pleurozium schreberi*
knight's plume moss								▮	▮	*Ptilium crista-castrensis*

Approximate cover classes (%): ▮ <2   ▬ 2–4.99   ▬ 5–9.99   ▬ 10–19.99   ▬ ≥20.

Nutrient Regime

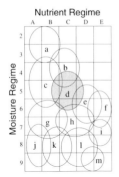

CHARACTERISTIC SPECIES

Tree	[54]	white spruce
	[3]	balsam fir[x]
	[2]	aspen*
	[1]	white birch*
	[1]	balsam poplar*
Shrub	[8]	balsam fir
	[4]	twin-flower
	[3]	green alder*
	[2]	low-bush cranberry*
	[1]	prickly rose
Forb	[4]	wild sarsaparilla
	[4]	bunchberry
	[2]	dewberry
	[1]	wild lily-of-the-valley
Moss	[35]	stair-step moss
	[23]	Schreber's moss
	[9]	knight's plume moss

SITE CHARACTERISTICS

Moisture Regime: mesic[6], submesic[2], subhygric[2]
Nutrient Regime: medium[7], poor[2], rich[1]
Topographic Position: midslope[5], level[2], upper slope[1], lower slope[1]

Stand Age: 91.9 ± 3.9 yrs; n = 21
Richness: 22.84 ± 1.31; n = 31
Diversity: 1.87 ± 0.11; n = 31

SOIL CHARACTERISTICS

Organic Thickness: (6–15)[7], (0–5)[2]
Humus Form: mor[8], raw moder[2]
Surface Texture: Si[2], SiS[2], SiL[2], SL[1], LmS[1]
Effective Texture: C[3], SC[1], SiS[1], SiL[1], SiC[1], SCL[1]
Depth to Mottles/Gley: none[9], (0–25)[1]
Drainage: well[5], mod. well[4], rapid[1]
Depth to Water Table: none[10]
Parent Material: M[4], GL[2], FL[1], GL/M[1], F[1]
Soil Subgroup: O.GL[4], BR.GL[4], E.EB[1]
Soil Association: Do[2], Bt[2], Wv[1], Wt[1], Pr[1], Ln[1]
Soil Type: SM4[7], SD2[1]

PLANT COMMUNITY TYPES (n)

d4.1 wS/green alder (4)
d4.2 wS/balsam fir/feather moss (11)
d4.3 wS/feather moss (16)

---

* Species characteristic of the phase but occurring in <70% of the sample plots with a prominence value <20.
[x] Tree may be dominant in some plots.

d4 low-bush cranberry wS - Plant community types and approximate cover classes

		d4.1	d4.2	d4.3
TREE	white spruce — *Picea glauca*			
	balsam fir — *Abies balsamea*			
	aspen — *Populus tremuloides*			
SHRUB	balsam fir — *Abies balsamea*			
	green alder — *Alnus crispa*			
	low-bush cranberry — *Viburnum edule*			
	twin-flower — *Linnaea borealis*			
	prickly rose — *Rosa acicularis*			
FORB	wild sarsaparilla — *Aralia nudicaulis*			
	bunchberry — *Cornus canadensis*			
	dewberry — *Rubus pubescens*			
MOSS	stair-step moss — *Hylocomium splendens*			
	Schreber's moss — *Pleurozium schreberi*			
	knight's plume moss — *Ptilium crista-castrensis*			

Approximate cover classes (%): ■ <2 ■ 2–4.99 ■ 5–9.99 ■ 10–19.99 ■ ≥20.

GENERAL DESCRIPTION

The dogwood ecosite is subhygric and nutrient-rich. These sites are commonly found in midslope or lower slope topographic positions or near watercourses where they receive nutrient-rich seepage or floodwaters for a portion of the growing season. Fine-textured morainal and glaciolacustrine parent materials are common and plant communities tend to be high in species richness, cover, and diversity.

e1 dogwood bP-tA   **d3 low-bush cranberry tA-wS**

SUCCESSIONAL RELATIONSHIPS

Succession proceeds slowly after disturbance due to the proliferation of grass, forb, and shrub cover. This explosion of vegetational cover can make tree establishment (especially coniferous) difficult and can reduce early tree species growth rates. Once white spruce becomes established, high growth rates can be expected.

MANAGEMENT INTERPRETATIONS

Drought Limitations	Excess Moisture	Rutting Hazard	Compaction Hazard	Puddling Hazard	Erosion Hazard	Frost Heave Hazard	Soil Temperature Limitations	Vegetation Competition	Windthrow Hazard	Productivity	Season of Harvest
L	M-H	H	H	H	H	H	M	H	L-M	H	W

subhygric/rich

Nutrient Regime

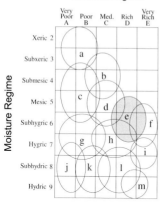

INDICATOR SPECIES

balsam poplar
bracted honeysuckle
dogwood, currants
wild red raspberry, tall lungwort
sweet-scented bedstraw
oak fern, lady fern, shield fern

PRODUCTIVITY

Site Index at 50 years BH age:
white spruce: 18.5 ± 0.5 m; n = 19
balsam poplar: 20.1 ± 0.6 m; n = 17
aspen: 21.3 ± 0.5 m; n = 16
white birch: 18.8 ± 1.7 m; n = 6
balsam fir: 16.0 ± 1.2 m; n = 5
black spruce: 13.6 ± 1.1 m; n = 2

Mean Annual Increment:
3.4 ± 0.2 m³/ha/yr; n = 33

SITE CHARACTERISTICS

Moisture Regime: subhygric[4], mesic[3], hygric[2]
Nutrient Regime: rich[7], medium[3]
Topographic Position: midslope[3], lower slope[2], level[2], depression[1], upper slope[1]

SOIL CHARACTERISTICS

Organic Thickness: (6–15)[7], (16–25)[1], (0–5)[1]
Humus Form: mor[5], raw moder[3], moder[1]
Surface Texture: SiL[3], SL[1], C[1]
Effective Texture: C[3], SiC[1], SiCL[1], CL[1], hC[1], SiL[1]
Depth to Mottles/Gley: none[4], (0–25)[3], (51–100)[1], (26–50)[1]
Drainage: mod. well[4], imperfect[3], well[2], poor[1]
Depth to Water Table: none[8], (51–100)[1]
Parent Material: M[3], GL[1], F[1]
Soil Subgroup: O.GL[1], BR.GL[1], O.LG[1], R.HG[1]
Soil Association: Av[2], Ln[2], Wv[1], Aw[1]
Soil Type: SM4[5], SWm[2]

ECOSITE PHASES (n)

e1 dogwood bP-tA (29)
e2 dogwood bP-wS (21)
e3 dogwood wS (15)

---

e1 | dogwood bP-tA (n = 29)

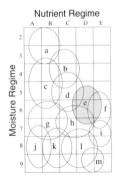

Nutrient Regime

Moisture Regime

SITE CHARACTERISTICS

Moisture Regime: subhygric[5], mesic[2], hygric[2]
Nutrient Regime: rich[8], medium[1]
Topographic Position: midslope[3], level[3], lower slope[2], depression[1], toe[1]

Stand Age: 72.5 ± 4.5 yrs; n = 15
Richness: 34.72 ± 1.87; n = 29
Diversity: 3.34 ± 0.09; n = 29

SOIL CHARACTERISTICS

Organic Thickness: (6–15)[7], (16–25)[1], (0–5)[1]
Humus Form: mor[5], raw moder[3], moder[2]
Surface Texture: SiL[3], C[1], Si[1], SL[1]
Effective Texture: C[3], hC[2], SiC[2], CL[1]
Depth to Mottles/Gley: (0–25)[4], none[4], (26–50)[1]
Drainage: imperfect[4], mod. well[3], well[1], poor[1]
Depth to Water Table: none[9], (51–100)[1]
Parent Material: M[3], GL[1], F[1]
Soil Subgroup: O.LG[1], O.GL[1], CU.R[1], BR.GL[1]
Soil Association: Av[2], Do[2], Wv[1], Wt[1], Ln[1], Kw[1], Hw[1], Aw[1]
Soil Type: SM4[5], SWm[2]

CHARACTERISTIC SPECIES

Tree
[32] balsam poplar
[23] aspen
[5] white birch[x]
[2] white spruce*

Shrub
[10] river alder
[10] dogwood
[6] low-bush cranberry
[5] prickly rose
[5] wild red raspberry
[4] currant
[3] alder-leaved buckthorn*
[2] bracted honeysuckle*
[1] balsam poplar*
[1] mountain maple*
[1] high-bush cranberry*

Forb
[10] wild sarsaparilla
[3] bunchberry
[3] dewberry
[3] fireweed
[1] tall lungwort
[1] palmate-leaved coltsfoot
[1] bishop's-cap
[1] sweet-scented bedstraw
[+] oak fern*
[+] shield fern*

PLANT COMMUNITY TYPES (n)

e1.1 bP-tA/dogwood/fern (10)
e1.2 bP-tA/bracted honeysuckle/
 fern (4)
e1.3 bP-tA/river alder–green alder/
 fern (13)
e1.4 bP-tA/alder-leaved buckthorn (1)
e1.5 bP-tA/mountain maple (1)

---

* Species characteristic of the phase but occurring in <70% of the sample plots with a prominence value <20.
[x] Tree may be dominant in some plots.

e1 dogwood bP-tA - Plant community types and approximate cover classes

		e1.1	e1.2	e1.3	e1.4	e1.5	
TREE	balsam poplar	▮▮▮	▮▮▮	▮▮▮	▮▮	▮▮	Populus balsamifera
	aspen		▮	▮▮		▮▮	Populus tremuloides
	white birch		▮	▮▮	▮	▮▮	Betula papyrifera
SHRUB	alder-leaved buckthorn	▮		▮▮	▮▮		Rhamnus alnifolia
	dogwood	▮	▮	▮▮▮	▮▮	▮▮▮	Cornus stolonifera
	mountain maple			▮			Acer spicatum
	river alder	▮		▮	▮		Alnus tenuifolia
	prickly rose	▮	▮	▮▮	▮		Rosa acicularis
	green alder	▮▮					Alnus crispa
	wild red raspberry	▮▮	▮	▮	▮		Rubus idaeus
	low-bush cranberry	▮	▮	▮			Viburnum edule
	willow					▮	Salix spp.
	currant	▮	▮	▮	▮	▮▮	Ribes spp.
	pin and choke cherry					▮	Prunus spp.
	white birch						Betula papyrifera
	bracted honeysuckle					▮	Lonicera involucrata
	balsam poplar						Populus balsamifera
	high-bush cranberry	·					Viburnum opulus*
FORB	wild sarsaparilla	▮	▮	▮▮▮	▮	▮	Aralia nudicaulis
	late goldenrod	▮	▮	▮▮▮	▮▮		Solidago gigantea
	fireweed	▮	▮	▮▮▮	▮▮		Epilobium angustifolium
	woodland horsetail	▮					Equisetum sylvaticum
	dewberry	▮	▮	▮▮▮	▮		Rubus pubescens
	bunchberry	▮				▮	Cornus canadensis
	lady fern	·	·	·	·		Athyrium filix-femina*
	common horsetail						Equisetum arvense
	tall lungwort	·	·	·	·	·	Mertensia paniculata*
	oak fern	·					Gymnocarpium dryopteris*
	shield fern	·					Dryopteris carthusiana*
GRASS	marsh reed grass	▮▮	▮	▮▮▮	▮▮▮		Calamagrostis canadensis

Approximate cover classes (%): · <2 ▮ 2-4.99 ▮ 5-9.99 ▮ 10-19.99 ▮ ≥20.

* Characteristic species with a prominence value <20 for all plant community types.

Nutrient Regime

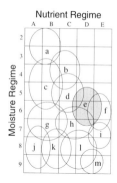

CHARACTERISTIC SPECIES

Tree
- [30] white spruce
- [12] aspen
- [10] white birch
- [9] balsam poplar
- [3] balsam fir*

Shrub
- [7] balsam fir
- [5] river alder
- [5] prickly rose
- [5] mountain maple*
- [4] dogwood
- [4] bracted honeysuckle
- [3] wild red raspberry
- [2] currant
- [2] low-bush cranberry
- [1] alder-leaved buckthorn*
- [+] bush honeysuckle*
- [+] high-bush cranberry*

Forb
- [7] wild sarsaparilla
- [3] bunchberry
- [3] dewberry
- [3] bishop's-cap
- [2] tall lungwort
- [1] northern starflower
- [1] oak fern*

Moss
- [8] stair-step moss
- [5] Schreber's moss

SITE CHARACTERISTICS

Moisture Regime: subhygric[4], mesic[3], hygric[3], subhydric[1]
Nutrient Regime: rich[7], medium[4]
Topographic Position: midslope[4], upper slope[2], lower slope[2], depression[2]

Stand Age: 79.1 ± 6.5 yrs; n = 13
Richness: 32.29 ± 1.52; n = 21
Diversity: 3.24 ± 0.13; n = 21

SOIL CHARACTERISTICS

Organic Thickness: (6–15)[7], (0–5)[1]
Humus Form: mor[5], raw moder[3], peatymor[1]
Surface Texture: SiL[3], cS[1], LmS[1]
Effective Texture: C[3], SiC[1], cS[1], SiCL[1]
Depth to Mottles/Gley: none[5], (0–25)[2], (51–100)[2]
Drainage: imperfect[4], mod. well[3], well[2], poor[2]
Depth to Water Table: none[8], (0–25)[1]
Parent Material: M[4], L[2], GL[1]
Soil Subgroup: O.LG[2], BR.GL[2], O.GL[1]
Soil Association: Ln[3], Bx[2], Wv[1], Av[1]
Soil Type: SM4[5], SWm[3], SM1[1]

PLANT COMMUNITY TYPES (n)

- e2.1 bP-wS/dogwood/fern (6)
- e2.2 bP-wS/bracted honeysuckle/ fern (3)
- e2.3 bP-wS/river alder–green alder/ fern (3)
- e2.4 bP-wS/bush honeysuckle (1)
- e2.5 bP-wS/mountain maple (2)
- e2.6 bP-wS/balsam fir/fern (3)
- e2.7 bP-wS/fern/feather moss (3)

---

* Species characteristic of the phase but occurring in <70% of the sample plots with a prominence value <20.

e2 dogwood bP-wS

Plant community types on following pages

e2 dogwood bP-wS - Plant community types and approximate cover classes

	Common name	Scientific name	e2.1	e2.2	e2.3	e2.4	e2.5	e2.6	e2.7
TREE	white spruce	*Picea glauca*	■	■	■	■	■	■	■
	white birch	*Betula papyrifera*	■	■	■		■	■	■
	balsam poplar	*Populus balsamifera*	■	■	■	■		■	
	aspen	*Populus tremuloides*		■		■		■	
	mountain maple	*Acer spicatum*					■	■	■
	balsam fir	*Abies balsamea*			■		■	■	■
SHRUB	mountain maple	*Acer spicatum*			■	■	■	■	
	balsam fir	*Abies balsamea*				■		■	
	green alder	*Alnus crispa*			■				
	bracted honeysuckle	*Lonicera involucrata*		■	■	■			
	wild red raspberry	*Rubus idaeus*		■	■			■	
	river alder	*Alnus tenuifolia*		■	■	■			
	dogwood	*Cornus stolonifera*		■					
	prickly rose	*Rosa acicularis*	■	■		■			
	bush honeysuckle	*Diervilla lonicera*			■	■			
	alder-leaved buckthorn	*Rhamnus alnifolia*			■				
	beaked hazelnut	*Corylus cornuta*				■			
	currant	*Ribes spp.*				■			
	low-bush cranberry	*Viburnum edule*	■	■	■	■			
	snowberry	*Symphoricarpos albus*							
	bearberry	*Arctostaphylos uva-ursi*	■	■	■				■
	white birch	*Betula papyrifera*						■	
FORB	wild sarsaparilla	*Aralia nudicaulis*	■	■	■	■	■	■	■
	bunchberry	*Cornus canadensis*	■	■	■	■	■	■	■
	dewberry	*Rubus pubescens*	■	■	■	■	■	■	
	bishop's-cap	*Mitella nuda*				■	■	■	
	dwarf scouring rush	*Equisetum scirpoides*					■		
	tall lungwort	*Mertensia paniculata*				■	■		
	kidney-leaved violet	*Viola renifolia*				■	■		
	meadow horsetail	*Equisetum pratense*					■		
	stiff club-moss	*Lycopodium annotinum*			■	■		■	

e2 dogwood bP-wS - Plant community types and approximate cover classes (concluded)

	e2.1	e2.2	e2.3	e2.4	e2.5	e2.6	e2.7	
FORB (continued)								
oak fern	■				■			*Gymnocarpium dryopteris**
Lindley's aster				▬				*Aster ciliolatus*
common pink wintergreen				▬				*Pyrola asarifolia*
palmate-leaved coltsfoot		▬			■	▮		*Petasites palmatus*
shield fern	■		▬	■			▮	*Dryopteris carthusiana**
GRASS								
marsh reed grass			▬			▮		*Calamagrostis canadensis*
sedge				▬				*Carex* spp.
millet grass								*Milium effusum*
MOSS								
stair-step moss		■				▰		*Hylocomium splendens*
Schreber's moss			▬■			▰		*Pleurozium schreberi*
knight's plume moss						▮		*Ptilium crista-castrensis*
mnium moss	▬					■		*Mnium* spp.

Approximate cover classes (%): ■ <2   ▬ 2–4.99   ▰ 5–9.99   ▰ 10–19.99   ▰ ≥20.
* Characteristic species with a prominence value <20 for all plant community types.

e3 dogwood wS (n = 15)

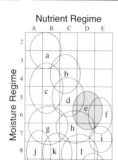

Nutrient Regime

Moisture Regime

CHARACTERISTIC SPECIES

Tree
[30] white spruce
[21] balsam fir
[2] aspen*
[2] white birch*
[1] balsam poplar*

Shrub
[10] balsam fir
[4] low-bush cranberry
[4] river alder
[2] prickly rose
[2] currant*
[2] dogwood*

Forb
[7] wild sarsaparilla
[6] bunchberry
[5] woodland horsetail
[5] dewberry
[3] bishop's-cap
[3] tall lungwort
[1] sweet-scented bedstraw
[1] kidney-leaved violet
[1] oak fern*
[1] wild lily-of-the-valley
[+] shield fern*

Moss
[22] stair-step moss
[14] Schreber's moss
[6] knight's plume moss

SITE CHARACTERISTICS

Moisture Regime: mesic[3], subhygric[3], hygric[2], subhydric[1]
Nutrient Regime: rich[6], medium[3]
Topographic Position: midslope[3], lower slope[2], upper slope[1], level[1], depression[1]

Stand Age: 119.7 ± 13.7 yrs; n = 8
Richness: 30.00 ± 1.91; n = 15
Diversity: 2.65 ± 0.22; n = 15

SOIL CHARACTERISTICS

Organic Thickness: (6–15)[6], (16–25)[2]
Humus Form: mor[5], raw moder[3], peatymor[2], moder[1]
Surface Texture: SiL[4], mesic[2], SL[2], SiCL[1], LmS[1], L[1]
Effective Texture: C[3], mesic[2], SiL[2], SiCL[2], CL[2], SL[1]
Depth to Mottles/Gley: none[5], (0–25)[3], (26–50)[1]
Drainage: mod. well[5], well[2], poor[1], imperfect[1]
Depth to Water Table: none[6], (51–100)[1], (26–50)[1], (0–25)[1]
Parent Material: M[3], GL/M[1]
Soil Subgroup: O.GL[3], T.M[1], R.HG[1], D.GL[1], BR.GL[1]
Soil Association: Lc[3], Av[3], Fs[1], Fh[1], Bt[1], Aw[1]
Soil Type: SM4[3], SWm[2], SR[2], SWp[1], SM3[1], SM2[1], SD4[1]

PLANT COMMUNITY TYPES (n)

e3.1 wS/river alder/fern (3)
e3.2 wS/balsam fir/fern (8)
e3.3 wS/fern/feather moss (4)

---

* Species characteristic of the phase but occurring in <70% of the sample plots with a prominence value <20.

e3 dogwood wS - Plant community types and approximate cover classes

		e3.1	e3.2	e3.3
TREE	white spruce	▬	▬	
	balsam fir	▪	▮	▮
	white birch *Betula papyrifera* *	▬	▪	▪
SHRUB	river alder *Alnus tenuifolia*	▬		
	balsam fir *Abies balsamea*	▪	▪	
	low-bush cranberry *Viburnum edule*	▪	▪	
	currant *Ribes* spp.	▪		
	white birch *Betula papyrifera* *	▪	▪	
	prickly rose *Rosa acicularis*	▪		▪
FORB	woodland horsetail *Equisetum sylvaticum*	▬	▪	▬
	wild sarsaparilla *Aralia nudicaulis*	▪		
	bunchberry *Cornus canadensis*	▪	▪	
	dwarf scouring rush *Equisetum scirpoides*	▪	▪	
	dewberry *Rubus pubescens*	▪	▪	
	bishop's-cap *Mitella nuda*	▪	▪	
	tall lungwort *Mertensia paniculata*		▪	
	oak fern *Gymnocarpium dryopteris* *		▪	▪
	shield fern *Dryopteris carthusiana* *	▪	▪	▪
GRASS	sedge *Carex* spp.	▬	▬	
MOSS	stair-step moss *Hylocomium splendens*	▪	▪	▪
	Schreber's moss *Pleurozium schreberi*	▪	▪	▪
	common tree moss *Climacium dendroides*	▪		
	knight's plume moss *Ptilium crista-castrensis*	▪	▪	▪
	brown moss *Drepanocladus* spp.		▪	

Approximate cover classes (%): ▪ <2   ▪ 2–4.99   ▬ 5–9.99   ▬ 10–19.99   ▬ ≥20.
* Characteristic species with a prominence value <20 for all plant community types.

GENERAL DESCRIPTION

The ostrich fern ecosite (f) is subhygric and nutrient-rich. It frequently occurs on fluvial parent materials where periodic flooding replenishes the nutrients available for plants. It also occurs where freshwater springs come to the surface and in lower slope areas where seepage is prevalent. In approximately 90% of the plots sampled mottles or gley occurred within 25 cm of the soil surface which is indicative of the high moisture availability. This ecosite tends to be high in species richness, cover, and diversity and is uncommon in the Mid-Boreal Upland Ecoregion. It is however, the dominant ecosite along watercourses of the Mid-Boreal Lowland Ecoregion. The ostrich fern mM-wE-bP-gA ecosite phase (f2) occurs exclusively in the Mid-Boreal Lowland Ecoregion and is the only ecological unit with Manitoba maple (*Acer negundo*), white elm (*Ulmus americana*), and green ash (*Fraxinus pennsylvanica*). The ostrich fern ecosite tends to be the most productive in the Mid-Boreal ecoregions of Saskatchewan.

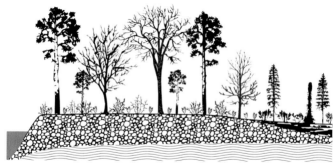

f2 ostrich fern mM-wE-bP-gA　　　　**k1 treed poor fen**

SUCCESSIONAL RELATIONSHIPS

Stands composed of aspen, balsam poplar, white birch, Manitoba maple, white elm, and/or green ash are successional to white spruce-dominated sites. If white spruce does not colonize this ecosite after disturbance it may have difficulties becoming established due to excessive competition and/or frequent disturbance by flooding in fluvial environments. Historically, white spruce trees were common in areas consisting of the ostrich fern ecosite, but most large spruce have been harvested. After the selective harvest of the white spruce, its regeneration has been inadequate. This may be due to the use of improper or limited silvicultural practices.

MANAGEMENT INTERPRETATIONS

Drought Limitations	Excess Moisture	Rutting Hazard	Compaction Hazard	Puddling Hazard	Erosion Hazard	Frost Heave Hazard	Soil Temperature Limitations	Vegetation Competition	Windthrow Hazard	Productivity	Season of Harvest
L	M-H	H	H	H	H	H	M	H	L-M	H	W

subhygric/very rich

Nutrient Regime

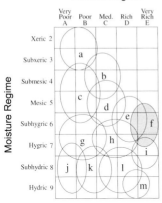

INDICATOR SPECIES

balsam poplar
high-bush cranberry
dogwood
currants
ostrich fern

PRODUCTIVITY

Site Index at 50 years BH age:
white spruce: 23.6 ± 2.0 m; n = 5
balsam poplar: 22.3 ± 1.0 m; n = 4
Manitoba maple: 16.1 ± 2.1 m; n = 4
white elm: 15.8 ± 0.8 m; n = 3
aspen: 21.4 ± 0.8 m; n = 2
white birch: 20.6 m; n = 1

Mean Annual Increment:
3.1 ± 0.5 m³/ha/yr; n = 9

SITE CHARACTERISTICS

Moisture Regime: subhygric[7], hygric[2], mesic[1]
Nutrient Regime: rich[9], very rich[1]
Topographic Position: level[5], upper slope[2], lower slope[2], depression[1]

SOIL CHARACTERISTICS

Organic Thickness: (6–15)[9]
Humus Form: mor[7], raw moder[2], moder[1]
Surface Texture: SiL[4], SiCL[3], SiC[1]
Effective Texture: SiCL[4], SiL[3], SiC[2]
Depth to Mottles/Gley: (0–25)[9], none[1]
Drainage: imperfect[6], mod. well[2], poor[1]
Depth to Water Table: none[9]
Parent Material: F[8], GL[1]
Soil Subgroup: R.G[2], GLCU.R[2], GL.R[2]
Soil Association: Sk[5], Mw[2], Av[2]
Soil Type: SM4[6], SWm[2], SM3[2]

ECOSITE PHASES

f1 ostrich fern bP-tA (5)
f2 ostrich fern mM-wE-bP-gA (10)
f3 ostrich fern bP-wS (2)

f1 | ostrich fern bP-tA (n = 5)

Nutrient Regime

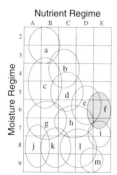

CHARACTERISTIC SPECIES

Tree
[19] balsam poplar
[19] white birch
[6] aspen
[+] Manitoba maple*

Shrub
[40] mountain maple
[12] beaked hazelnut
[7] high-bush cranberry
[5] dogwood
[5] currant
[1] pin and choke cherry

Forb
[14] ostrich fern
[11] wild sarsaparilla
[4] meadow horsetail
[4] dewberry
[3] bishop's-cap
[1] tall lungwort
[1] Lindley's aster
[1] lady fern*
[1] sweet-scented bedstraw
[1] northern bedstraw
[+] red and white baneberry
[+] common pink wintergreen

Grass
[3] sedge

Moss
[+] ragged moss

SITE CHARACTERISTICS

Moisture Regime: subhygric[4], hygric[4], mesic[2]
Nutrient Regime: rich[8], very rich[2]
Topographic Position: upper slope[4], lower slope[4], midslope[2]

Stand Age: no data
Richness: 32.60 ± 2.71; n = 5
Diversity: 2.92 ± 0.23; n = 5

SOIL CHARACTERISTICS

Organic Thickness: (6–15)[8], (0–5)[2]
Humus Form: mor[10]
Surface Texture: SiL[7], SL[3]
Effective Texture: SiL[3], SiC[3], C[3]
Depth to Mottles/Gley: (0–25)[7], none[3]
Drainage: mod. well[4], well[2], poor[2], imperfect[2]
Depth to Water Table: none[10]
Parent Material: GL[4], M/GL[2], M[2], F[2]
Soil Subgroup: O.HG[2], HU.LG[2], GL.R[2], CU.R[2], BR.GL[2]
Soil Association: Ln[5] Aw[5]
Soil Type: SWm[7], SM4[3]

PLANT COMMUNITY TYPES (n)

f1.1 bP-tA/dogwood/ostrich fern (2)
f1.2 bP-tA/mountain maple/ ostrich fern (3)

---

* Species characteristic of the phase but occurring in <70% of the sample plots with a prominence value <20.

f1 ostrich fern bP-tA - Plant community types and approximate cover classes

		f1.1	f1.2
TREE			
balsam poplar	*Populus balsamifera*		
white birch	*Betula papyrifera*		
aspen	*Populus tremuloides*		
SHRUB			
mountain maple	*Acer spicatum*		
beaked hazelnut	*Corylus cornuta*		
high-bush cranberry	*Viburnum opulus*		
dogwood	*Cornus stolonifera*		
currant	*Ribes* spp.		
aspen	*Populus tremuloides*		
saskatoon	*Amelanchier alnifolia*		
prickly rose	*Rosa acicularis*		
low-bush cranberry	*Viburnum edule* *		
FORB			
ostrich fern	*Matteuccia struthiopteris*		
wild sarsaparilla	*Aralia nudicaulis*		
meadow horsetail	*Equisetum pratense*		
dewberry	*Rubus pubescens*		
bishop's-cap	*Mitella nuda*		
kidney-leaved violet	*Viola renifolia*		
lady fern	*Athyrium filix-femina* *		
tall lungwort	*Mertensia paniculata* *		
GRASS			
sedge	*Carex* spp.		
MOSS			
common beaked moss	*Eurhynchium pulchellum*		

Approximate cover classes (%): ■ <2 ■ 2–4.99 ■ 5–9.99 ■ 10–19.99 ■ ≥20.

* Characteristic species with a prominence value <20 for all plant community types.

f2 · ostrich fern mM-wE-bP-gA (n = 10)

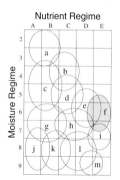

Nutrient Regime
Moisture Regime

CHARACTERISTIC SPECIES

Tree

[30] Manitoba maple
[14] balsam poplar
[13] white elm
[8] white birch
[6] green ash
[+] white spruce*

Shrub

[13] Manitoba maple
[9] pin and choke cherry
[8] high-bush cranberry
[5] green ash
[3] currant
[2] river alder*
[2] dogwood
[1] white elm
[1] balsam poplar
[+] mountain maple*

Forb

[27] ostrich fern
[7] meadow horsetail
[4] bishop's-cap
[3] wild sarsaparilla
[1] dewberry
[1] great-spurred violet
[1] small enchanter's nightshade*
[1] sweet-scented bedstraw
[+] red and white baneberry
[+] shield fern*

Grass

[2] sedge

SITE CHARACTERISTICS

Moisture Regime: subhygric[8], mesic[1], hygric[1]
Nutrient Regime: rich[10]
Topographic Position: level[8], depression[1], upper slope[1]

Stand age: 71.4 ± 11.1 yrs; n = 8
Richness: 26.10 ± 1.77; n = 10
Diversity: 3.17 ± 0.18; n = 10

SOIL CHARACTERISTICS

Organic Thickness: (6–15)[10]
Humus Form: mor[6], raw moder[2], moder[2]
Surface Texture: SiCL[4], SiL[3], hC[1], SiC[1]
Effective Texture: SiCL[6], SiL[2], hC[1], SiC[1]
Depth to Mottles/Gley: (0–25)[9], none[1]
Drainage: imperfect[7], mod. well[2], poor[1]
Depth to Water Table: none[9], (51–100)[1]
Parent Material: F[10]
Soil Subgroup: R.G[3], GLCU.R[3], GL.R[2], R.HG[1], O.HR[1]
Soil Association: Sk[8], Mw[2]
Soil Type: SM4[7], SM3[2], SWm[1]

PLANT COMMUNITY TYPES (n)

f2.1 mM-wE-bP-gA/pin cherry–saskatoon/ostrich fern (4)
f2.2 mM-wE-bP-gA/Manitoba maple/ostrich fern (4)
f2.3 mM-wE-bP-gA/high-bush cranberry–green ash/ostrich fern (2)

* Species characteristic of the phase but occurring in <70% of the sample plots with a prominence value <20.

f2 ostrich fern mM-wE-bP-gA - Plant community types and approximate cover classes

		f2.1	f2.2	f2.3
TREE	Manitoba maple — *Acer negundo*			
	balsam poplar — *Populus balsamifera*			
	white birch — *Betula papyrifera*			
	white elm — *Ulmus americana*			
	green ash — *Fraxinus pennsylvanica*			
	aspen — *Populus tremuloides*			
SHRUB	Manitoba maple — *Acer negundo*			
	green ash — *Fraxinus pennsylvanica*			
	pin and choke cherry — *Prunus* spp.			
	high-bush cranberry — *Viburnum opulus*			
	currant — *Ribes* spp.			
	river alder — *Alnus tenuifolia*			
	willow — *Salix* spp.			
	dogwood — *Cornus stolonifera*			
	wild red raspberry — *Rubus idaeus*			
	white elm — *Ulmus americana* *			
FORB	ostrich fern — *Matteuccia struthiopteris*			
	meadow horsetail — *Equisetum pratense*			
	bishop's-cap — *Mitella nuda*			
	wild sarsaparilla — *Aralia nudicaulis*			
	great-spurred violet — *Viola selkirkii*			
	dewberry — *Rubus pubescens* *			
GRASS	sedge — *Carex* spp.			

Approximate cover classes (%): ■ <2   ■ 2–4.99   ■ 5–9.99   ■ 10–19.99   ■ ≥20.
* Characteristic species with a prominence value <20 for all plant community types.

f3 | ostrich fern bP-wS (n = 2)

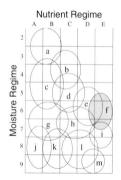

Nutrient Regime

Moisture Regime

CHARACTERISTIC SPECIES

Tree
[19] white spruce
[12] balsam poplar
[7] balsam fir[x]
[6] white birch[x]
[5] aspen*

Shrub
[16] dogwood
[5] prickly rose
[5] balsam fir
[5] bracted honeysuckle
[4] currant
[4] alder-leaved buckthorn
[3] river alder
[2] low-bush cranberry
[2] wild red raspberry

Forb
[14] ostrich fern
[10] wild sarsaparilla
[4] tall lungwort
[1] bunchberry
[1] Lindley's aster
[1] wild lily-of-the-valley
[1] dewberry
[1] woodland strawberry
[1] northern bedstraw
[1] sweet-scented bedstraw
[1] bishop's-cap
[1] kidney-leaved violet

Moss
[2] leafy moss*
[1] ragged moss

SITE CHARACTERISTICS

Moisture Regime: subhygric[10]
Nutrient Regime: very rich[5], rich[5]
Topographic Position: depression[5], lower slope[5]

Stand Age: 103.0 yrs; n = 1
Richness: 38.00 ± 5.00; n = 2
Diversity: 3.89 ± 0.35; n = 2

SOIL CHARACTERISTICS

Organic Thickness: (6–15)[10]
Humus Form: raw moder[5], mor[5]
Surface Texture: SiL[5], SiC[5]
Effective Texture: SiL[5], SiC[5]
Depth to Mottles/Gley: (0–25)[10]
Drainage: imperfect[10]
Depth to Water Table: none[10]
Parent Material: F[10]
Soil Subgroup: R.G[5], GL.HR[5]
Soil Association: Av[10]
Soil Type: SM4[5], SM3[5]

PLANT COMMUNITY TYPES (n)

f3.1 bP-wS/dogwood/ostrich fern (2)

---

* Species characteristic of the phase but occurring in <70% of the sample plots with a prominence value <20.
[x] Tree may be dominant in some plots.

f3 ostrich fern bP-wS - Plant community types and approximate cover classes

f3.1

TREE

white spruce	*Picea glauca*
balsam poplar	*Populus balsamifera*
balsam fir	*Abies balsamea*
white birch	*Betula papyrifera*
aspen	*Populus tremuloides*

SHRUB

dogwood	*Cornus stolonifera*
prickly rose	*Rosa acicularis*
balsam fir	*Abies balsamea*
bracted honeysuckle	*Lonicera involucrata*
currant	*Ribes* spp.
alder-leaved buckthorn	*Rhamnus alnifolia*
river alder	*Alnus tenuifolia*
low-bush cranberry	*Viburnum edule*
wild red raspberry	*Rubus idaeus* *

FORB

ostrich fern	*Matteuccia struthiopteris*
wild sarsaparilla	*Aralia nudicaulis*
tall lungwort	*Mertensia paniculata*
bunchberry	*Cornus canadensis* *
Lindley's aster	*Aster ciliolatus* *
wild lily-of-the-valley	*Maianthemum canadense* *
dewberry	*Rubus pubescens* *
woodland strawberry	*Fragaria vesca* *
northern bedstraw	*Galium boreale* *
sweet-scented bedstraw	*Galium triflorum* *
bishop's-cap	*Mitella nuda* *
kidney-leaved violet	*Viola renifolia* *

MOSS

leafy moss	*Plagiomnium* spp. *
ragged moss	*Brachythecium* spp. *

Approximate cover classes (%): ■ <2 ■ 2–4.99 ■ 5–9.99 ■ 10–19.99 ■ ≥20.
* Characteristic species with a prominence value <20 for all plant community types.

g | Labrador tea–hygric (n = 32)

GENERAL DESCRIPTION

The Labrador tea–hygric ecosite has a nutrient-poor substrate with imperfectly to poorly drained soils. Labrador tea and bog cranberry are indicative of the relatively acidic surface soil conditions. It dominantly occurs on fine-textured till or glaciolacustrine parent material where the wet soil conditions promote the development of Gleysolic soils. While the Labrador tea–hygric ecosite has plant community types similar to the Labrador tea–submesic ecosite (c), the hygric ecosite tends to occur in lower topographic positions, has mottles in the top 25 cm of soil, has a thicker organic layer, and may be dominated by black spruce rather than pine. The high soil water content associated with this ecosite creates a greater risk of site modification when the soils are not frozen. This ecosite is not common in the Mid-Boreal Lowland Ecoregion of Saskatchewan.

c1 Labrador tea–submesic jP-bS   g1 Labrador tea–hygric bS-jP

SUCCESSIONAL RELATIONSHIPS

Young and mature stands developing in this ecosite often have a component of black spruce. The black spruce is often the same age as the pine but forms a secondary canopy due to slower growth rates. Successionally mature stands are dominated by black spruce with a small component of old residual pine.

MANAGEMENT INTERPRETATIONS

Drought Limitations	Excess Moisture	Rutting Hazard	Compaction Hazard	Puddling Hazard	Erosion Hazard	Frost Heave Hazard	Soil Temperature Limitations	Vegetation Competition	Windthrow Hazard	Productivity	Season of Harvest
L	M-H	H	H	H	H	H	H	M	H	L-M	W

hygric/poor

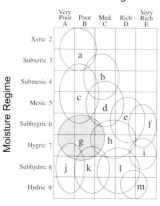

Nutrient Regime

INDICATOR SPECIES

black spruce
jack pine
Labrador tea
bog cranberry
feather moss

PRODUCTIVITY

Site Index at 50 years BH age:
black spruce: 12.9 ± 0.7 m; n = 23
jack pine: 15.5 ± 1.0 m; n = 6
white spruce: 10.9 m; n = 1

Mean Annual Increment:
2.0 ± 0.2 m³/ha/yr; n = 25

SITE CHARACTERISTICS

Moisture Regime: subhygric[4], hygric[3], subhydric[2]
Nutrient Regime: poor[6], medium[3]
Topographic Position: level[3], lower slope[3], midslope[2]

SOIL CHARACTERISTICS

Organic Thickness: (6–15)[5], (40–59)[1], (26–39)[1], (16–25)[1], (0–5)[1]
Humus Form: mor[6], raw moder[3]
Surface Texture: Si[2], mS[1], LmS[1], SCL[1]
Effective Texture: C[3], SC[1], LmS[1]
Depth to Mottles/Gley: (0–25)[8]
Drainage: imperfect[6], poor[3]
Depth to Water Table: none[6], (51–100)[2], >100[1]
Parent Material: M[3], GL[2], GF/M[1], F[1]
Soil Subgroup: O.LG[4], GL.GL[2], R.G[1], O.G[1]
Soil Association: Ln[3], Aw[2], Bt[1], Lc[1], Do[1], Bo[1], Av[1]
Soil Type: SWp[3], SWm[2], SM4[2]

ECOSITE PHASES (n)

g1 Labrador tea–hygric bS-jP (32)

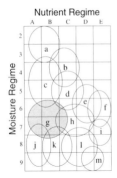

Nutrient Regime

Moisture Regime

CHARACTERISTIC SPECIES

Tree
[42] black spruce
[5] jack pine[x]

Shrub
[12] Labrador tea
[2] blueberry*
[2] black spruce*
[2] bog cranberry
[1] prickly rose*
[1] twin-flower*

Forb
[1] bunchberry*
[1] woodland horsetail*
[1] dwarf scouring rush*
[1] palmate-leaved coltsfoot*

Moss
[47] Schreber's moss
[26] stair-step moss
[10] knight's plume moss
[+] peat moss

Lichen
[3] reindeer lichen

SITE CHARACTERISTICS

Moisture Regime: subhygric[4], hygric[3], subhydric[2]
Nutrient Regime: poor[6], medium[3]
Topographic Position: level[3], lower slope[3], midslope[2]

Stand age: 92.7 ± 5.1 yrs; n = 27
Richness: 16.63 ± 1.70; n = 32
Diversity: 1.50 ± 1.03; n = 32

SOIL CHARACTERISTICS

Organic Thickness: (6–15)[5], (40–59)[1], (26–39)[1], (16–25)[1], (0–5)[1]
Humus Form: mor[6], raw moder[3]
Surface Texture: Si[2], mS[1], LmS[1], SCL[1]
Effective Texture: C[3], SC[1], LmS[1]
Depth to Mottles/Gley: (0–25)[8]
Drainage: imperfect[6], poor[3]
Depth to Water Table: none[6], (51–100)[2], >100[1]
Parent Material: M[3], GL[2], GF/M[1], F[1]
Soil Subgroup: O.LG[4], GL.GL[2], R.G[1], O.G[1]
Soil Association: Ln[3], Aw[2], Bt[1], Lc[1], Do[1], Bo[1], Av[1]
Soil Type: SWp[3], SWm[2], SM4[2]

PLANT COMMUNITY TYPES (n)

g1.1 bS-jP/Labrador tea/feather moss (14)
g1.2 bS-jP/feather moss (18)

---

* Species characteristic of the phase but occurring in <70% of the sample plots with a prominence value <20.
[x] Tree may be dominant in some plots.

g1 Labrador tea bS-jP - Plant community types and approximate cover classes

		g1.1	g1.2	
TREE	black spruce	▪	▮	*Picea mariana*
	jack pine		▮	*Pinus banksiana* *
	white spruce		▮	*Picea glauca*
SHRUB	Labrador tea	▮		*Ledum groenlandicum*
	black spruce	▪	•	*Picea mariana*
	blueberry	▪	•	*Vaccinium myrtilloides* *
	bog cranberry	▪	•	*Vaccinium vitis-idaea* *
	prickly rose	▪	▪	*Rosa acicularis* *
MOSS	Schreber's moss	▮	▮	*Pleurozium schreberi*
	stair-step moss	▪	▮	*Hylocomium splendens*
	knight's plume moss	▪	▮	*Ptilium crista-castrensis*
	peat moss	▪	▪	*Sphagnum* spp. *
LICHEN	reindeer lichen	▪		*Cladina* spp.

Approximate cover classes (%): ▪ <2   ▪ 2–4.99   ▪ 5–9.99   ▮ 10–19.99   ▮ ≥20.
* Characteristic species with a prominence value <20 for all plant community types.

| h | **horsetail** (n = 28) |

GENERAL DESCRIPTION

The horsetail ecosite is subhygric to hygric and nutrient rich. These sites are commonly found on fluvial or glaciolacustrine parent materials where flooding or seepage enhances the substrate nutrient supply. With high water tables, wet soil conditions, and Gleysolic soils, organic matter tends to accumulate. Horsetails commonly form a blanket over the forest floor.

d4 low-bush cranberry wS   **h3 horsetail wS-bS**

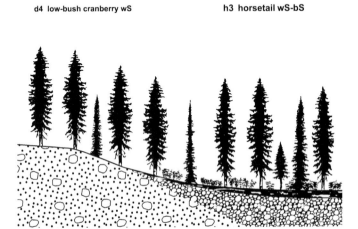

SUCCESSIONAL RELATIONSHIPS

Succession on these sites is largely controlled by high soil water content. Some sites that have peaty soils may have taken hundreds of years to develop. When the trees are removed, the water table may rise making tree establishment difficult; understory vegetation development is aggressive following disturbance (e.g., *Calamagrostis canadensis*). White spruce forms the canopy in the last successional stage.

MANAGEMENT INTERPRETATIONS

Drought Limitations	Excess Moisture	Rutting Hazard	Compaction Hazard	Puddling Hazard	Erosion Hazard	Frost Heave Hazard	Soil Temperature Limitations	Vegetation Competition	Windthrow Hazard	Productivity	Season of Harvest
L	H	H	H	H	H	H	H	H	H	H	W

hygric/rich | h

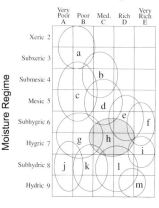

Nutrient Regime

(Moisture Regime vs Nutrient Regime diagram with labeled ellipses a–m; columns: Very Poor A, Poor B, Med. C, Rich D, Very Rich E; rows: Xeric 2, Subxeric 3, Submesic 4, Mesic 5, Subhygric 6, Hygric 7, Subhydric 8, Hydric 9)

INDICATOR SPECIES

common horsetail
meadow horsetail

PRODUCTIVITY

Site Index at 50 years BH age:
white spruce: 18.1 ± 1.0 m; n = 14
balsam poplar: 15.0 ± 0.6 m; n = 4
black spruce: 14.3 ± 0.9 m; n = 4
aspen: 19.9 ± 1.4 m; n = 4
tamarack: 23.0 m; n = 1

Mean Annual Increment:
3.3 ± 0.2 m³/ha/yr; n = 13

SITE CHARACTERISTICS

Moisture Regime: subhygric[5], hygric[5]
Nutrient Regime: rich[7], medium[3]
Topographic Position: level[3], lower slope[3], depression[2], toe[1]

SOIL CHARACTERISTICS

Organic Thickness: (6–15)[4], (16–25)[3], (26–39)[1], (0–5)[1]
Humus Form: mor[4], peatymor[3], raw moder[1], moder[1]
Surface Texture: Si[2], SiC[2], SiL[1], LmS[1]
Effective Texture: SiC[3], humic[1], Si[1]
Depth to Mottles/Gley: (0–25)[6], none[1]
Drainage: poor[5], imperfect[4]
Depth to Water Table: none[4], >100[2], (51–100)[2], (26–50)[2], (0–25)[2]
Parent Material: F[3], GL[2], L[1]
Soil Subgroup: R.G[2], R.HG[1], O.HG[1], T.H[1], O.G[1]
Soil Association: Av[4], Fb[2], Aw[2]
Soil Type: SM4[3], SWp[2], SWm[2], SR[2], SM3[1]

ECOSITE PHASES (n)

h1 horsetail bP-tA (3)
h2 horsetail bP-wS (10)
h3 horsetail wS-bS (15)

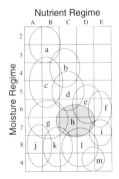

Nutrient Regime

Moisture Regime

SITE CHARACTERISTICS

Moisture Regime: subhygric[10]
Nutrient Regime: rich[10]
Topographic Position: toe[7], lower slope[3]

Stand Age: 56.0 yrs; n = 1
Richness: 34.67 ± 7.06; n = 3
Diversity: 3.70 ± 0.09; n = 3

SOIL CHARACTERISTICS

Organic Thickness: (6–15)[3], (16–25)[3], (0–5)[3]
Humus Form: raw moder[10]
Surface Texture: SiL[3], SiC[3], LmS[3]
Effective Texture: Si[3], SCL[3], C[3]
Depth to Mottles/Gley: (0–25)[7], none[3]
Drainage: imperfect[3], poor[3], well[3]
Depth to Water Table: none[10]
Parent Material: F[7], C[3]
Soil Subgroup: R.G[7], O.EB[3]
Soil Association: Av[10]
Soil Type: SM4[7], SM3[3]

CHARACTERISTIC SPECIES

Tree
[43] aspen
[18] balsam poplar
[1] white spruce*

Shrub
[19] willow
[12] wild red raspberry
[11] prickly rose
[6] currant
[3] twin-flower
[2] dogwood*
[2] white spruce*
[1] aspen

Forb
[17] meadow horsetail
[14] common horsetail
[13] wild sarsaparilla
[7] tall lungwort
[5] bunchberry
[5] wild lily-of-the-valley
[3] wild strawberry
[3] Lindley's aster
[3] fireweed
[2] northern bedstraw
[1] red and white baneberry

Grass
[11] marsh reed grass

PLANT COMMUNITY TYPES (n)

h1.1 bP-tA/horsetail (3)

---

* Species characteristic of the phase but occurring in <70% of the sample plots with a prominence value <20.

---

h1 horsetail bP-tA - Plant community types and approximate cover classes

		h1.1	
TREE			
	aspen	▮	*Populus tremuloides*
	balsam poplar	▮	*Populus balsamifera*
	white spruce	▪	*Picea glauca**
SHRUB			
	willow	▮	*Salix* spp.
	wild red raspberry	▮	*Rubus idaeus*
	prickly rose	▮	*Rosa acicularis*
	currant	▮	*Ribes* spp.
	twin-flower	▮	*Linnaea borealis*
	dogwood	▪	*Cornus stolonifera**
	white spruce	▪	*Picea glauca**
	aspen	▪	*Populus tremuloides**
FORB			
	meadow horsetail	▮	*Equisetum pratense*
	common horsetail	▮	*Equisetum arvense*
	wild sarsaparilla	▮	*Aralia nudicaulis*
	tall lungwort	▮	*Mertensia paniculata*
	bunchberry	▮	*Cornus canadensis*
	wild lily-of-the-valley	▪	*Maianthemum canadense*
	wild strawberry	▪	*Fragaria virginiana*
	Lindley's aster	▪	*Aster ciliolatus*
	fireweed	▪	*Epilobium angustifolium*
	northern bedstraw	▪	*Galium boreale**
	red and white baneberry	▪	*Actaea rubra**
GRASS			
	marsh reed grass	▮	*Calamagrostis canadensis*

Approximate cover classes (%): ▪ <2 ▮ 2–4.99 ▮ 5–9.99 ▮ 10–19.99 ▮ ≥20.
* Characteristic species with a prominence value <20 for all plant community types.

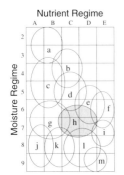

Nutrient Regime

Moisture Regime

CHARACTERISTIC SPECIES

Tree
- [30] white spruce
- [11] balsam poplar
- [11] white birch
- [7] aspen
- [5] balsam fir

Shrub
- [5] balsam fir
- [5] bracted honeysuckle
- [4] prickly rose
- [3] low-bush cranberry
- [3] white spruce
- [2] white birch
- [2] currant
- [2] twin-flower

Forb
- [16] meadow horsetail
- [6] common horsetail
- [4] woodland horsetail
- [4] wild sarsaparilla
- [4] bunchberry
- [4] dewberry
- [3] tall lungwort
- [2] bishop's-cap
- [2] palmate-leaved coltsfoot
- [1] common pink wintergreen

Grass
- [1] marsh reed grass

Moss
- [12] Schreber's moss
- [11] stair-step moss
- [6] knight's plume moss

SITE CHARACTERISTICS

Moisture Regime: subhygric[5], hygric[4], subhydric[1]
Nutrient Regime: rich[7], medium[3]
Topographic Position: level[4], depression[2], toe[2], lower slope[2]

Stand Age: 106.0 ± 7.5 yrs; n = 4
Richness: 33.80 ± 3.52; n = 10
Diversity: 3.40 ± 0.17; n = 10

SOIL CHARACTERISTICS

Organic Thickness: (6–15)[5], (16–25)[3], ≥80[1], (0–5)[1]
Humus Form: mor[4], raw moder[2], peatymor[2], moder[2]
Surface Texture: SiC[2], Si[2], mS[2], L[2], CL[1], mesic[1]
Effective Texture: SiC[3], Si[2], SCL[1], SC[1], L[1], CL[1], mesic[1]
Depth to Mottles/Gley: (0–25)[7], (51–100)[1], (26–50)[1]
Drainage: imperfect[6], poor[4]
Depth to Water Table: none[4], >100[2], (51–100)[2], (0–25)[2]
Parent Material: F[4], L[2], S[1], GL/GF[1], GL[1], GF/M[1]
Soil Subgroup: O.G[3], R.HG[1], R.G[1], O.HG[1], GLD.GL[1], GL.HR[1], GL.EB[1], CU.M[1]
Soil Association: Av[6], Lc[2], Fb[2]
Soil Type: SWm[4], SM4[2], SM3[2], SR[1], SMp[1]

PLANT COMMUNITY TYPES (n)

h2.1 bP-wS/horsetail (10)

h2 horsetail bP-wS - Plant community types and approximate cover classes

	h2.1
TREE	
white spruce	*Picea glauca*
balsam poplar	*Populus balsamifera*
white birch	*Betula papyrifera*
aspen	*Populus tremuloides*
balsam fir	*Abies balsamea*
SHRUB	
balsam fir	*Abies balsamea*
bracted honeysuckle	*Lonicera involucrata*
prickly rose	*Rosa acicularis*
low-bush cranberry	*Viburnum edule*
white spruce	*Picea glauca*
white birch	*Betula papyrifera**
currant	*Ribes* spp.*
twin-flower	*Linnaea borealis**
FORB	
meadow horsetail	*Equisetum pratense*
common horsetail	*Equisetum arvense*
woodland horsetail	*Equisetum sylvaticum*
wild sarsaparilla	*Aralia nudicaulis*
bunchberry	*Cornus canadensis*
dewberry	*Rubus pubescens*
tall lungwort	*Mertensia paniculata*
bishop's-cap	*Mitella nuda*
palmate-leaved coltsfoot	*Petasites palmatus**
common pink wintergreen	*Pyrola asarifolia**
GRASS	
marsh reed grass	*Calamagrostis canadensis**
MOSS	
Schreber's moss	*Pleurozium schreberi*
stair-step moss	*Hylocomium splendens*
knight's plume moss	*Ptilium crista-castrensis*

Approximate cover classes (%): ■ <2   ■ 2-4.99   ■ 5-9.99   ■ 10-19.99   ▬ ≥20.
* Characteristic species with a prominence value <20 for all plant community types.

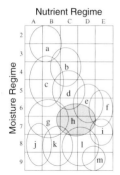

Nutrient Regime

Moisture Regime

CHARACTERISTIC SPECIES

Tree

[39] white spruce
[4] black spruce[x]
[1] balsam poplar*
[1] white birch*
[1] aspen*

Shrub

[7] Labrador tea
[4] twin-flower
[2] low-bush cranberry
[1] currant
[1] prickly rose

Forb

[23] meadow horsetail
[11] common horsetail
[5] bunchberry
[4] dewberry
[3] palmate-leaved coltsfoot
[2] tall lungwort
[2] bishop's-cap

Grass

[5] marsh reed grass

Moss

[22] Schreber's moss
[19] stair-step moss
[9] knight's plume moss

SITE CHARACTERISTICS

Moisture Regime: hygric[6],
subhygric[3]
Nutrient Regime: rich[7], medium[3]
Topographic Position: level[4],
lower slope[3], depression[2]

Stand Age: 105.8 ± 7.5 yrs; n = 10
Richness: 30.07 ± 2.80; n = 15
Diversity: 2.61 ± 0.14; n = 15

SOIL CHARACTERISTICS

Organic Thickness: (6–15)[4],
(26–39)[3], (16–25)[3]
Humus Form: mor[5], peatymor[4],
moder[1]
Surface Texture: Si[3], fibric[2], SiL[2],
LmS[2]
Effective Texture: SiC[4], humic[2],
hC[2], SiL[2]
Depth to Mottles/Gley: (0–25)[5],
none[2]
Drainage: poor[6], imperfect[2],
mod. well[1]
Depth to Water Table: none[3],
(26–50)[3], >100[1], (51–100)[1], (0–25)[1]
Parent Material: GL[3]
Soil Subgroup: T.H[2], R.HG[2],
O.HG[2], R.G[1]
Soil Association: Aw[4], Fb[3], Do[1], Av[1]
Soil Type: SWp[5], SR[2], SM4[2]

PLANT COMMUNITY TYPES (n)

h3.1 wS-bS/horsetail (8)
h3.2 wS-bS/Labrador tea/horsetail (7)

---

* Species characteristic of the phase but occurring in <70% of the sample plots with a prominence value <20.
[x] Tree may be dominant in some plots.

h3 horsetail wS-bS - Plant community types and approximate cover classes

		h3.1	h3.2	
TREE	white spruce	▮		*Picea glauca*
	black spruce		▮	*Picea mariana*
	tamarack		▮	*Larix laricina*
	balsam poplar	▪	▪	*Populus balsamifera* *
	white birch	▪	▪	*Betula papyrifera* *
	aspen		▪	*Populus tremuloides* *
SHRUB	Labrador tea	▪	▮	*Ledum groenlandicum*
	twin-flower	▪	▮	*Linnaea borealis*
	low-bush cranberry	▪		*Viburnum edule*
FORB	meadow horsetail	▮	▮	*Equisetum pratense*
	common horsetail	▮	▮	*Equisetum arvense*
	bunchberry	▮	▪	*Cornus canadensis*
	palmate-leaved coltsfoot	▪		*Petasites palmatus*
	dewberry	▪	▪	*Rubus pubescens*
	tall lungwort	▪		*Mertensia paniculata*
GRASS	marsh reed grass	▮		*Calamagrostis canadensis*
MOSS	Schreber's moss	▮	▮	*Pleurozium schreberi*
	stair-step moss	▮	▮	*Hylocomium splendens*
	knight's plume moss	▮	▪	*Ptilium crista-castrensis*

Approximate cover classes (%): ▪ <2   ▪ 2–4.99   ▪ 5–9.99   ▮ 10–19.99   ▬ ≥20.
* Characteristic species with a prominence value <20 for all plant community types.

GENERAL DESCRIPTION

The gully ecosite is a unique ecological unit that occurs in drainage swales or channels where springs, small streams, and runoff provide a continuous supply of water and nutrients. The wet soil conditions facilitate the accumulation of organic matter. Organic soils have developed over fluvial parent materials in nearly half of the plots sampled. Approximately 70% of the sample sites with mineral soils had gley or mottles within 25 cm of the mineral soil surface and all sample sites with mineral soils had gley or mottles within 100 cm of the soil surface.

d3 low-bush cranberry
    tA-wS

i1 river alder gully

d2 low-bush
    cranberry tA

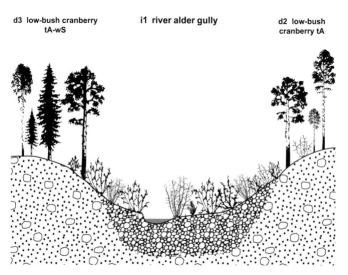

SUCCESSIONAL RELATIONSHIPS

Due to a combination of the excessively wet conditions and the relatively frequent fluvial disturbance regime, trees have difficulty becoming established on this ecosite. The river alder gully ecosite phase (i1) could be considered an edaphic climax. However, it is transitional to the ostrich fern ecosite (f) where white spruce forms the canopy of the climax community.

MANAGEMENT INTERPRETATIONS

Drought Limitations	Excess Moisture	Rutting Hazard	Compaction Hazard	Pudding Hazard	Erosion Hazard	Frost Heave Hazard	Soil Temperature Limitations	Vegetation Competition	Windthrow Hazard	Productivity	Season of Harvest
L	H	H	M	M	H	H	H	H	not applicable		

hygric/very rich

Nutrient Regime

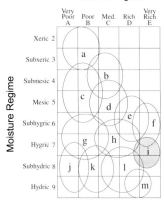

INDICATOR SPECIES

river alder
dogwood
willow
ostrich fern
oak fern
shield fern

PRODUCTIVITY

not applicable

SITE CHARACTERISTICS

Moisture Regime: subhydric[7], hygric[3]
Nutrient Regime: rich[8], very rich[2]
Topographic Position: depression[7], toe[2], level[2]

SOIL CHARACTERISTICS

Organic Thickness: ≥80[3], (6–15)[3], (26–39)[3]
Humus Form: peatymor[7], mull[2], mor[2]
Surface Texture: mS[2], humic[2], fibric[2], cS[2], SiL[2], CL[2]
Effective Texture: mS[2], humic[2], fibric[2], Si[2], cS[2], C[2]
Depth to Mottles/Gley: (0–25)[3], none[2], (51–100)[2]
Drainage: poor[10]
Depth to Water Table: (26–50)[3], (0–25)[3], >100[2], (51–100)[2]
Parent Material: F[5], S/GF[2], S/F[2], F/GL[2]
Soil Subgroup: T.H[2], T.F[2], R.G[2], O.HR[2], O.G[2], GL.HR[2]
Soil Association: Av[7], Fs[3]
Soil Type: SWm[3], SR[3], SWp[3]

ECOSITE PHASES (n)

i1 river alder gully (6)

Nutrient Regime

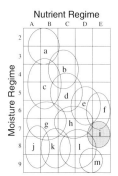

CHARACTERISTIC SPECIES

Shrub
[77] river alder
[10] dogwood
[8] currant
[6] wild red raspberry
[4] willow
[4] prickly rose
[3] bracted honeysuckle
[2] low-bush cranberry*
[2] balsam poplar

Forb
[14] ostrich fern
[6] woodland horsetail
[4] oak fern
[1] tall lungwort
[1] bishop's-cap
[1] dewberry
[1] shield fern*
[+] sweet-scented bedstraw

Grass
[4] sedge
[+] drooping wood-reed

Moss
[1] woodsy leafy moss*
[1] common tree moss*
[1] brown moss*

SITE CHARACTERISTICS

Moisture Regime: subhydric[7], hygric[3]
Nutrient Regime: rich[8], very rich[2]
Topographic Position: depression[7], toe[2], level[2]

Stand Age: no data
Richness: 35.66 ± 3.19; n = 6
Diversity: 2.80 ± 0.29; n = 6

SOIL CHARACTERISTICS

Organic Thickness: ≥80[3], (6–15)[3], (26–39)[3]
Humus Form: peatymor[7], mull[2], mor[2]
Surface Texture: mS[2], humic[2], fibric[2], cS[2], SiL[2], CL[2]
Effective Texture: mS[2], humic[2], fibric[2], Si[2], cS[2], C[2]
Depth to Mottles/Gley: (0–25)[3], none[2], (51–100)[2]
Drainage: poor[10]
Depth to Water Table: (26–50)[3], (0–25)[3], >100[2], (51–100)[2]
Parent Material: F[5], S/GF[2], S/F[2], F/GL[2]
Soil Subgroup: T.H[2], T.F[2], R.G[2], O.HR[2], O.G[2], GL.HR[2]
Soil Association: Av[7], Fs[3]
Soil Type: SWm[3], SR[3], SWp[3]

PLANT COMMUNITY TYPES (n)

i1.1 river alder/ostrich fern (6)

---

* Species characteristic of the phase but occurring in <70% of the sample plots with a prominence value <20.

i1 river alder gully - Plant community types and approximate cover classes

i1.1

SHRUB

		i1.1
river alder	*Alnus tenuifolia*	
dogwood	*Cornus stolonifera*	
currant	*Ribes* spp.	
wild red raspberry	*Rubus idaeus*	
willow	*Salix* spp.	
prickly rose	*Rosa acicularis*	
bracted honeysuckle	*Lonicera involucrata*	
low-bush cranberry	*Viburnum edule* *	
balsam poplar	*Populus balsamifera* *	

FORB

ostrich fern	*Matteuccia struthiopteris*	
woodland horsetail	*Equisetum sylvaticum*	
oak fern	*Gymnocarpium dryopteris*	
tall lungwort	*Mertensia paniculata* *	
bishop's-cap	*Mitella nuda* *	
dewberry	*Rubus pubescens* *	
shield fern	*Dryopteris carthusiana* *	
sweet-scented bedstraw	*Galium triflorum* *	

GRASS

sedge	*Carex* spp.	
drooping wood-reed	*Cinna latifolia* *	

MOSS

woodsy leafy moss	*Plagiomnium cuspidatum* *	
common tree moss	*Climacium dendroides* *	
brown moss	*Drepanocladus* spp. *	

Approximate cover classes (%): ■ <2 ■ 2–4.99 ■ 5–9.99 ■ 10–19.99 ■ ≥20.
* Characteristic species with a prominence value <20 for all plant community types.

GENERAL DESCRIPTION

The bog ecosite commonly has organic soils consisting of slowly decomposing peat moss. They are poorly to very poorly drained and have a very poor to poor nutrient regime. This ecosite occupies level and depressional areas where water tends to be stagnant and impeded drainage or high water tables enhance the accumulation of organic matter. Stunted black spruce form a sparse canopy on the treed phase (j1) of the bog ecosite. Along with the poor fen (k) and rich fen (l) ecosites the bog dominates the flat landscape of the Mid-Boreal Lowland Ecoregion.

j1 treed bog   **g1 Labrador tea–hygric bS-jP**

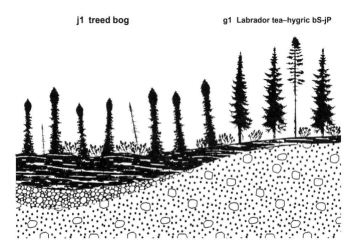

SUCCESSIONAL RELATIONSHIPS

The bog ecosite is an edaphic climax that is maintained by high water tables. The hydrarch succession to the bog ecosite is extremely slow.

MANAGEMENT INTERPRETATIONS

Drought Limitations	Excess Moisture	Rutting Hazard	Compaction Hazard	Puddling Hazard	Erosion Hazard	Frost Heave Hazard	Soil Temperature Limitations	Vegetation Competition	Windthrow Hazard	Productivity	Season of Harvest
L	H	H	L	L	L	H	H	L	H	L	not applicable

subhydric/very poor

Nutrient Regime

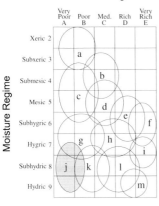

INDICATOR SPECIES

black spruce
Labrador tea
bog cranberry
cloudberry
peat moss
small bog cranberry
slender hair-cap

PRODUCTIVITY

Site Index at 50 years BH age:
black spruce: 8.0 ± 0.9 m; n = 13
tamarack: 12.3 ± 2.0 m; n = 3
balsam poplar: 20.0 m; n = 1
white spruce: 10.8 m; n = 1

Mean Annual Increment:
0.8 ± 0.2 m³/ha/yr; n = 8

SITE CHARACTERISTICS

Moisture Regime: subhydric[6], hydric[2], hygric[1], subhygric[1]
Nutrient Regime: very poor[6], poor[3]
Topographic Position: level[5], depression[5]

SOIL CHARACTERISTICS

Organic Thickness: ≥80[8], (16–25)[1]
Humus Form: peatymor[9], mor[1]
Surface Texture: fibric[8], mesic[1]
Effective Texture: fibric[6], mesic[2], humic[1]
Depth to Mottles/Gley: not applicable
Drainage: poor[5], very poor[4]
Depth to Water Table: (26–50)[5], (0–25)[2], >100[2], (51–100)[2]
Parent Material: O[6], O/M[1], M[1], B[1]
Soil Subgroup: T.M[3], TY.F[2], TME.F[1], O.LG[1]
Soil Association: Bf[4], Bb[3]
Soil Type: SR[9], SWp[1]

ECOSITE PHASES (n)

j1 treed bog (21)
j2 shrubby bog (6)

---

j1 | **treed bog** (n = 21)

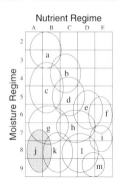

Nutrient Regime

Moisture Regime

CHARACTERISTIC SPECIES

Tree
[33] black spruce

Shrub
[49] Labrador tea
[11] black spruce
[7] bog cranberry
[1] small bog cranberry*

Forb
[9] cloudberry

Moss
[39] peat moss
[37] Schreber's moss
[3] stair-step moss
[1] knight's plume moss*
[1] slender hair-cap*

Lichen
[4] reindeer lichen

SITE CHARACTERISTICS

Moisture Regime: subhydric[6], hydric[2], hygric[2]
Nutrient Regime: very poor[6], poor[3]
Topographic Position: level[5], depression[5]

Stand Age: 121.8 ± 12.7 yrs; n = 10
Richness: 12.24 ± 1.31; n = 21
Diversity: 1.53 ± 0.08; n = 21

SOIL CHARACTERISTICS

Organic Thickness: ≥80[7], (16–25)[1]
Humus Form: peatymor[8], mor[2]
Surface Texture: fibric[7], mesic[2]
Effective Texture: fibric[4], mesic[3], humic[2], SCL[1], L[1]
Depth to Mottles/Gley: not applicable
Drainage: very poor[5], poor[5]
Depth to Water Table: (26–50)[6], (51–100)[1], (0–25)[1], >100[1]
Parent Material: O[7], M[1], B[1]
Soil Subgroup: T.M.[3], TY.F[2], O.LG[1]
Soil Association: Bb[4], Ln[1], Fh[1], Bs[1], Bf[1], Aw[1]
Soil Type: SR[8], SWp[1]

PLANT COMMUNITY TYPES (n)

j1.1 bS/Labrador tea/cloudberry/
peat moss (21)

---

* Species characteristic of the phase but occurring in <70% of the sample plots with a prominence value <20.

j1 treed bog - Plant community types and approximate cover classes

		j1.1
TREE	black spruce	*Picea mariana*
SHRUB	Labrador tea	*Ledum groenlandicum*
	black spruce	*Picea mariana*
	bog cranberry	*Vaccinium vitis-idaea*
	small bog cranberry	*Oxycoccus microcarpus**
FORB	cloudberry	*Rubus chamaemorus*
MOSS	peat moss	*Sphagnum* spp.
	Schreber's moss	*Pleurozium schreberi*
	stair-step moss	*Hylocomium splendens*
	knight's plume moss	*Ptilium crista-castrensis**
	slender hair-cap	*Polytrichum strictum**
LICHEN	reindeer lichen	*Cladina* spp.

Approximate cover classes (%): ■ <2 ■ 2–4.99 ■ 5–9.99 ■ 10–19.99 ■ ≥20.
* Characteristic species with a prominence value <20 for all plant community types.

j2 shrubby bog (n = 6)

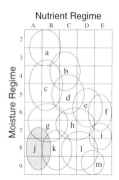

Nutrient Regime

Moisture Regime

CHARACTERISTIC SPECIES

Shrub
- [33] Labrador tea
- [22] black spruce
- [8] leatherleaf
- [5] bog cranberry
- [1] northern laurel
- [1] small bog cranberry

Forb
- [3] cloudberry

Moss
- [82] peat moss
- [1] Schreber's moss*
- [+] slender hair-cap*

Lichen
- [4] reindeer lichen
- [+] deformed cup lichen

SITE CHARACTERISTICS

Moisture Regime: subhydric[8], subhygric[2]
Nutrient Regime: very poor[7], poor[3]
Topographic Position: level[5], depression[5]

Stand Age: no data
Richness: 8.83 ± 0.40; n = 6
Diversity: 1.31 ± 0.06; n = 6

SOIL CHARACTERISTICS

Organic Thickness: ≥ 80[10]
Humus Form: peatymor[10]
Surface Texture: fibric[10]
Effective Texture: fibric[8], mesic[2]
Depth to Mottles/Gley: not applicable
Drainage: poor[7], very poor[3]
Depth to Water Table: $(0–25)$[4], >100[2], $(51–100)$[2], $(26–50)$[2]
Parent Material: O[5], O/M[3], O/GL[2]
Soil Subgroup: TME.F[3], TY.F[2], T.M[2], FI.OC[2], FI.M[2]
Soil Association: Bf[8], Bb[2]
Soil Type: SR[10]

PLANT COMMUNITY TYPES (n)

j2.1 black spruce–Labrador tea/ cloudberry/peat moss (6)

---

* Species characteristic of the phase but occurring in <70% of the sample plots with a prominence value <20.

j2 shrubby bog - Plant community types and approximate cover classes

		j2.1	
SHRUB	Labrador tea	▌	*Ledum groenlandicum*
	black spruce	▌	*Picea mariana*
	leatherleaf	▌	*Chamaedaphne calyculata*
	bog cranberry	▪	*Vaccinium vitis-idaea* *
	northern laurel	▪	*Kalmia polifolia* *
	small bog cranberry	▪	*Oxycoccus microcarpus* *
FORB	cloudberry	▪	*Rubus chamaemorus*
MOSS	peat moss	▌	*Sphagnum* spp.
	Schreber's moss	▌	*Pleurozium schreberi* *
	slender hair-cap	▪	*Polytrichum strictum* *
LICHEN	reindeer lichen	▪	*Cladina* spp.
	deformed cup lichen	▪	*Cladonia deformis* *

Approximate cover classes (%): ▪ <2   ▪ 2–4.99   ▬ 5–9.99   ▬ 10–19.99   ▬ ≥20.
* Characteristic species with a prominence value <20 for all plant community types.

GENERAL DESCRIPTION

The poor fen ecosite is intermediate in nutrient regime between the bog (j) and the rich fen (l) ecosites and as such has species characteristic of both. Drainage is poor to very poor, however, there is some movement of water through the substratum. This ecosite occupies level and depressional areas where impeded drainage or high water tables enhance the accumulation of organic matter. This organic matter consists of a combination of bog-type organic matter such as peat moss (*Sphagnum* spp.) and fen-type organic matter such as sedges (*Carex* spp.), golden moss (*Tomenthypnum nitens*), tufted moss (*Aulacomnium palustre*), and brown moss (*Drepanocladus* spp.). Both the black spruce and/or tamarack that dominate a sparse canopy on the treed phase (k1) of the poor fen ecosite are stunted and generally considered unmerchantable. Along with the bog (j) and rich fen (l) ecosites, the poor fen dominates the flat landscape of the Mid-Boreal Lowland Ecoregion.

k1 treed poor fen   **l2 shrubby rich fen**

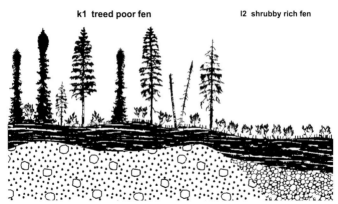

SUCCESSIONAL RELATIONSHIPS

The hydrarch succession characteristic of this ecosite occurs over a period of hundreds of years. Thus, recovery from disturbance is extremely slow. Site disturbance can influence the hydrologic regime, resulting in changes in the direction and rate of succession. As these systems depend on water flow through them, impeding this flow can result in a reduction or elimination of tree cover and changes in the shrub, forb, and grass layers.

MANAGEMENT INTERPRETATIONS

Drought Limitations	Excess Moisture	Rutting Hazard	Compaction Hazard	Puddling Hazard	Erosion Hazard	Frost Heave Hazard	Soil Temperature Limitations	Vegetation Competition	Windthrow Hazard	Productivity	Season of Harvest
L	H	H	L	L	L	H	H	L	H	L	not applicable

subhydric/medium

Nutrient Regime

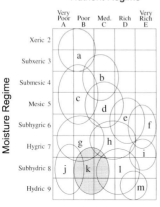

INDICATOR SPECIES

black spruce
tamarack
Labrador tea
dwarf birch
willow
cloudberry
sedge
peat moss, golden moss
tufted moss, brown moss

PRODUCTIVITY

Site Index at 50 years BH age:
tamarack: 13.2 ± 1.4 m; n = 10
black spruce: 10.7 ± 1.7 m; n = 6

Mean Annual Increment:
1.0 ± 0.1 m³/ha/yr; n = 8

SITE CHARACTERISTICS

Moisture Regime: subhydric[4], hydric[3], hygric[2]
Nutrient Regime: poor[5], medium[2], very rich[1], very poor[1]
Topographic Position: level[5], depression[4], toe[1]

SOIL CHARACTERISTICS

Organic Thickness: ≥80[7], (26–39)[2], (40–59)[1], (16–25)[1]
Humus Form: peatymor[8]
Surface Texture: fibric[7], mesic[1]
Effective Texture: fibric[6], mesic[2]
Depth to Mottles/Gley: (0–25)[1]
Drainage: very poor[7], poor[3]
Depth to Water Table: (0–25)[7], (51–100)[2]
Parent Material: O[5], B[1]
Soil Subgroup: TY.F[3], TY.M[2], T.F[1], T.M[1], O.HG[1]
Soil Association: Bf[3], Fh[2], Bb[2], Aw[2]
Soil Type: SR[9], SWp[1]

ECOSITE PHASES (n)

k1 treed poor fen (20)
k2 shrubby poor fen (9)

k1 treed poor fen (n = 20)

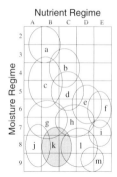

Nutrient Regime

Moisture Regime

CHARACTERISTIC SPECIES

Tree	[20]	black spruce
	[14]	tamarack
Shrub	[26]	Labrador tea
	[12]	black spruce
	[4]	bog cranberry
	[1]	small bog cranberry*
Forb	[+]	cloudberry*
Grass	[7]	sedge
Moss	[37]	peat moss
	[23]	Schreber's moss
	[6]	stair-step moss
	[3]	tufted moss
	[1]	slender hair-cap*
Lichen	[6]	reindeer lichen*

SITE CHARACTERISTICS

Moisture Regime: subhydric[5], hydric[3], hygric[2]
Nutrient Regime: poor[4], medium[3], very rich[1], very poor[1]
Topographic Position: level[5], depression[4], toe[1]

Stand Age: 84.1 ± 13.4 yrs; n = 9
Richness: 20.05 ± 1.72; n = 20
Diversity: 1.97 ± 0.09; n = 20

SOIL CHARACTERISTICS

Organic Thickness: ≥80[6], (26–39)[2], (40–59)[1]
Humus Form: peatymor[10]
Surface Texture: fibric[6], mesic[2], mS[1], humic[1], hC[1]
Effective Texture: fibric[5], mesic[3], mS[1], humic[1], SC[1]
Depth to Mottles/Gley: (0–25)[1]
Drainage: very poor[6], poor[4]
Depth to Water Table: (51–100)[5], (0–25)[5]
Parent Material: O[5], M[1], B/GF[1]
Soil Subgroup: TY.F[2], TY.M[2], T.M[1], T.F[1], O.HG[1]
Soil Association: Fh[3], Bb[3], Bf[2], Aw[2]
Soil Type: SR[9], SWp[1]

PLANT COMMUNITY TYPES (n)

k1.1 bS-tL/dwarf birch/sedge/peat moss (20)

---

* Species characteristic of the phase but occurring in <70% of the sample plots with a prominence value <20.

k1 treed poor fen - Plant community types and approximate cover classes

		k1.1	
TREE	black spruce	▮	*Picea mariana*
	tamarack	▮	*Larix laricina*
SHRUB	Labrador tea	▮	*Ledum groenlandicum*
	black spruce	▮	*Picea mariana*
	bog cranberry	▮	*Vaccinium vitis-idaea*
	small bog cranberry	▪	*Oxycoccus microcarpus* *
FORB	cloudberry	▪	*Rubus chamaemorus* *
GRASS	sedge	▮	*Carex* spp.
MOSS	peat moss	▮	*Sphagnum* spp.
	Schreber's moss	▮	*Pleurozium schreberi*
	stair-step moss	▮	*Hylocomium splendens*
	tufted moss	▮	*Aulacomnium palustre*
	slender hair-cap	▪	*Polytrichum strictum* *
LICHEN	reindeer lichen	▮	*Cladina* spp.

Approximate cover classes (%): ▪ <2 ▮ 2–4.99 ▬ 5–9.99 ▬ 10–19.99 ▬ ≥20.

* Characteristic species with a prominence value <20 for all plant community types.

k2 | shrubby poor fen (n = 9)

Nutrient Regime

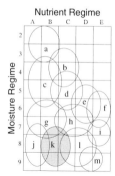

CHARACTERISTIC SPECIES

Shrub
- [17] leatherleaf
- [10] black spruce
- [9] dwarf birch
- [8] bog rosemary
- [6] tamarack
- [3] willow
- [2] Labrador tea*
- [1] small bog cranberry*

Grass
- [11] sedge
- [3] marsh reed grass*

Moss
- [71] peat moss
- [6] cushion moss
- [3] golden moss
- [1] tufted moss*
- [1] slender hair-cap*

Lichen
- [1] reindeer lichen*

SITE CHARACTERISTICS

Moisture Regime: hydric[4], subhydric[4], hygric[1]
Nutrient Regime: poor[8], very poor[1], very rich[1]
Topographic Position: depression[5], level[5]

Stand Age: no data
Richness: 15.78 ± 1.85; n = 9
Diversity: 1.55 ± 0.07; n = 9

SOIL CHARACTERISTICS

Organic Thickness: ≥80[9], (16–25)[1]
Humus Form: peatymor[9], mor[1]
Surface Texture: fibric[9], SiC[1]
Effective Texture: fibric[7], mesic[1], SiC[1]
Depth to Mottles/Gley: (0–25)[1]
Drainage: very poor[9], poor[1]
Depth to Water Table: (0–25)[9], (26–50)[1]
Parent Material: O[6], B[2], O/M[1], GL[1]
Soil Subgroup: TY.F[5], TY.M[3], T.F[1], GL.SC[1]
Soil Association: Bf[4], Fh[1], Fb[1], Bb[1], Aw[1]
Soil Type: SR[9], SWp[1]

PLANT COMMUNITY TYPES (n)

k2.1 black spruce–tamarack–dwarf birch/sedge/peat moss (9)

---

* Species characteristic of the phase but occurring in <70% of the sample plots with a prominence value <20.

ECOSITE PHASE

k2 shrubby poor fen - Plant community types and approximate cover classes

	k2.1
SHRUB	
leatherleaf	*Chamaedaphne calyculata*
black spruce	*Picea mariana*
dwarf birch	*Betula pumila*
bog rosemary	*Andromeda polifolia*
tamarack	*Larix laricina*
willow	*Salix* spp.
Labrador tea	*Ledum groenlandicum* *
small bog cranberry	*Oxycoccus microcarpus* *
GRASS	
sedge	*Carex* spp.
marsh reed grass	*Calamagrostis canadensis* *
MOSS	
peat moss	*Sphagnum* spp.
cushion moss	*Dicranum undulatum*
golden moss	*Tomenthypnum nitens* *
tufted moss	*Aulacomnium palustre* *
slender hair-cap	*Polytrichum strictum* *
LICHEN	
reindeer lichen	*Cladina* spp. *

Approximate cover classes (%): ■ <2   ■ 2–4.99   ■ 5–9.99   ■ 10–19.99   ■ ≥20.
* Characteristic species with a prominence value <20 for all plant community types.

I rich fen (n = 51)

GENERAL DESCRIPTION

The rich fen ecosite is characterized by flowing water and alkaline nutrient-rich conditions. The soil is composed of organic matter derived from decomposing sedges (*Carex* spp.), as well as golden moss (*Tomenthypnum nitens*), tufted moss (*Aulacomnium palustre*), and brown mosses (*Drepanocladus* spp.). This ecosite occupies level and depressional areas where the water table is at or near the surface for a portion of the growing season. Tamarack dominates the canopy in the treed phase while dwarf birch or willow form the canopy in the shrubby phase and sedges dominate the graminoid phase of the rich fen ecosite. Along with the bog (j) and poor fen (k) ecosites, the rich fen ecosite dominates the flat landscape of the Mid-Boreal Lowland Ecoregion.

l2 shrubby rich fen　　　　　　　　**d2 low-bush cranberry tA**

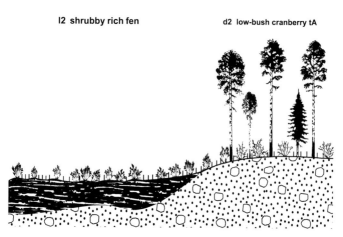

SUCCESSIONAL RELATIONSHIPS

The rich fen is an early stage in hydrarch succession. Species composition, direction, and rate of succession changes with the changing hydrologic regime. As with other wetlands, rich fens have slow successional rates so recovery from disturbance may also be slow.

MANAGEMENT INTERPRETATIONS

Drought Limitations	Excess Moisture	Rutting Hazard	Compaction Hazard	Puddling Hazard	Erosion Hazard	Frost Heave Hazard	Soil Temperature Limitations	Vegetation Competition	Windthrow Hazard	Productivity	Season of Harvest
L	H	H	L	L	L	H	H	L	H	L	not applicable

ECOSITE

subhydric/rich

Nutrient Regime

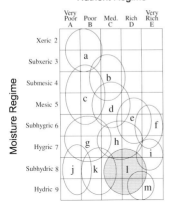

INDICATOR SPECIES

tamarack
willow
dwarf birch
sedge
golden moss
tufted moss
brown moss
marsh reed grass

PRODUCTIVITY

Site Index at 50 years BH age:
tamarack: 13.5 ± 2.0 m; n = 3
black spruce: 9.9 m; n = 1

Mean Annual Increment:
1.1 ± 0.1 m; n = 3

SITE CHARACTERISTICS

Moisture Regime: subhydric[5], hydric[3], hygric[2]
Nutrient Regime: medium[6], rich[3], poor[1]
Topographic Position: level[5], depression[4]

SOIL CHARACTERISTICS

Organic Thickness: ≥80[6], (40–59)[1]
Humus Form: peatymor[8], mor[1]
Surface Texture: fibric[6], mesic[1]
Effective Texture: fibric[5], mesic[2]
Depth to Mottles/Gley: (0–25)[2]
Drainage: very poor[6], poor[4]
Depth to Water Table: (0–25)[5], (26–50)[2], none[2]
Parent Material: O[3], N[2], O/M[1]
Soil Subgroup: TY.F[3], T.M[1], R.G[1]
Soil Association: Fh[4], Fs[2], Fb[2], Mw[1], Aw[1]
Soil Type: SR[8], SWp[1]

ECOSITE PHASES (n)

l1 treed rich fen (9)
l2 shrubby rich fen (22)
l3 graminoid rich fen (20)

Nutrient Regime

CHARACTERISTIC SPECIES

Tree

[23] tamarack

Shrub

[16] dwarf birch
[11] willow
[6] Labrador tea
[6] northern laurel
[6] river alder
[6] tamarack
[1] black spruce*

Forb

[4] buck-bean*
[3] marsh marigold
[2] three-leaved Solomon's seal
[1] marsh cinquefoil*

Grass

[16] sedge
[5] marsh reed grass

Moss

[17] peat moss
[13] tufted moss
[9] Schreber's moss
[6] golden moss
[2] brown moss*

SITE CHARACTERISTICS

Moisture Regime: hydric[4], subhydric[3], hygric[1], subhygric[1]
Nutrient Regime: medium[6], rich[3], very rich[1]
Topographic Position: depression[7], level[3]

Stand Age: 71.8 ± 19.0 yrs; n = 4
Richness: 21.67 ± 3.21; n = 9
Diversity: 2.55 ± 0.16; n = 9

SOIL CHARACTERISTICS

Organic Thickness: ≥80[8], (40–59)[2]
Humus Form: peatymor[10]
Surface Texture: fibric[7], mesic[2], mS[2]
Effective Texture: fibric[5], mesic[3], SCL[2]
Depth to Mottles/Gley: (26–50)[1]
Drainage: very poor[6], poor[4]
Depth to Water Table: (26–50)[4], (0–25)[4], (51–100)[2]
Parent Material: O[7], O/M[2], O/GF[1]
Soil Subgroup: TY.M[3], TY.F[3], T.F[1], T.M[1], R.G[1], FI.M[1]
Soil Association: Fs[4], Fh[2], Fb[2], Aw[2]
Soil Type: SR[9], SWp[1]

PLANT COMMUNITY TYPES (n)

I1.1 tL/dwarf birch/sedge/brown moss (9)

---

* Species characteristic of the phase but occurring in <70% of the sample plots with a prominence value <20.

I1 treed rich fen - Plant community types and approximate cover classes

		11.1
TREE	tamarack	*Larix laricina*
SHRUB	dwarf birch	*Betula pumila*
	willow	*Salix* spp.
	Labrador tea	*Ledum groenlandicum*
	northern laurel	*Kalmia polifolia*
	river alder	*Alnus tenuifolia*
	tamarack	*Larix laricina*
	black spruce	*Picea mariana* *
FORB	buck-bean	*Menyanthes trifoliata*
	marsh marigold	*Caltha palustris*
	three-leaved Solomon's seal	*Smilacina trifolia*
	marsh cinquefoil	*Potentilla palustris* *
GRASS	sedge	*Carex* spp.
	marsh reed grass	*Calamagrostis canadensis*
MOSS	peat moss	*Sphagnum* spp.
	tufted moss	*Aulacomnium palustre*
	Schreber's moss	*Pleurozium schreberi*
	golden moss	*Tomenthypnum nitens*
	brown moss	*Drepanocladus* spp.

Approximate cover classes (%): ■ <2   ■ 2–4.99   ■ 5–9.99   ■ 10–19.99   ■ ≥20.

* Characteristic species with a prominence value <20 for all plant community types.

Nutrient Regime

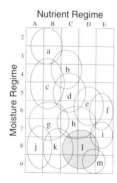

CHARACTERISTIC SPECIES

Shrub
[31] willow
[16] dwarf birch

Forb
[2] marsh cinquefoil

Grass
[37] sedge
[10] marsh reed grass
[4] fowl bluegrass

Moss
[10] brown moss
[5] tufted moss
[1] peat moss*

SITE CHARACTERISTICS

Moisture Regime: subhydric[6], hydric[2], hygric[1]
Nutrient Regime: medium[6], rich[3]
Topographic Position: level[6], depression[3]

Stand Age: no data
Richness: 18.14 ± 1.67; n = 22
Diversity: 2.25 ± 0.13; n = 22

SOIL CHARACTERISTICS

Organic Thickness: ≥80[6], (60–79)[1], (40–59)[1], (26–39)[1]
Humus Form: peatymor[8], mor[1]
Surface Texture: fibric[7], mesic[1]
Effective Texture: fibric[4], mesic[3]
Depth to Mottles/Gley: (0–25)[1]
Drainage: very poor[6], poor[3]
Depth to Water Table: (0–25)[5], (26–50)[3], none[2]
Parent Material: O[3], N[3], O/M[2], O/GF[1]
Soil Subgroup: TY.F[3], T.M[2], R.HG[1], FI.M[1]
Soil Association: Fh[3], Fs[3], Fb[2], Mw[1], Fp[1], Aw[1]
Soil Type: SR[8], SWp[1]

PLANT COMMUNITY TYPES (n)

I2.1 dwarf birch/sedge/golden moss (8)
I2.2 willow/sedge/golden moss (7)
I2.3 willow/marsh reed grass (7)

---

* Species characteristic of the phase but occurring in <70% of the sample plots with a prominence value <20.

I2 shrubby rich fen - Plant community types and approximate cover classes

		12.1	12.2	12.3	
SHRUB	willow	▪	■		*Salix* spp.
	dwarf birch	▪			*Betula pumila*
	bog rosemary	▪			*Andromeda polifolia*
	wild red raspberry			▪	*Rubus idaeus*
	tamarack	▪		▪	*Larix laricina*
	currant			▪	*Ribes* spp.
FORB	Canada thistle		■	▪	*Cirsium arvense*
	marsh cinquefoil			■	*Potentilla palustris*
	marsh marigold			■	*Caltha palustris*
	common nettle			■	*Urtica dioica*
	tall lungwort			■	*Mertensia paniculata*
GRASS	sedge	▪		▮	*Carex* spp.
	marsh reed grass			▮	*Calamagrostis canadensis*
	fowl bluegrass			■	*Poa palustris*
	fringed brome			■	*Bromus ciliatus*
MOSS	brown moss	■	■		*Drepanocladus* spp.
	golden moss	■			*Tomenthypnum nitens*
	tufted moss	■			*Aulacomnium palustre*
	fire moss		▪.	▪	*Ceratodon purpureus*
	peat moss	▪			*Sphagnum* spp. *

Approximate cover classes (%): ▪ <2, ■ 2–4.99, ■ 5–9.99, ▬ 10–19.99, ▬ ≥20.
* Characteristic species with a prominence value <20 for all plant community types.

| I3 | **graminoid rich fen** (n = 20)

Nutrient Regime

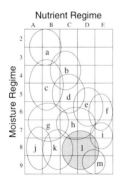

CHARACTERISTIC SPECIES

Shrub
[1] willow

Forb
[8] buck-bean

Grass
[38] sedge
[8] marsh reed grass

SITE CHARACTERISTICS

Moisture Regime: subhydric[4], hydric[4], hygric[2]
Nutrient Regime: medium[6], rich[2], poor[2]
Topographic Position: level[5], depression[4]

Stand Age: no data
Richness: 14.20 ± 1.47; n = 20
Diversity: 1.55 ± 0.18; n = 20

SOIL CHARACTERISTICS

Organic Thickness: ≥80[6], (6–15)[1], (40–59)[1], (16–25)[1]
Humus Form: peatymor[8], mor[1]
Surface Texture: fibric[6], C[1]
Effective Texture: fibric[6], humic[1], C[1]
Depth to Mottles/Gley: (0–25)[3]
Drainage: very poor[6], poor[4]
Depth to Water Table: (0–25)[6], none[2]
Parent Material: N[3], O/M[2], O[2], M[1]
Soil Subgroup: TY.F[3], R.G[1], HY.F[1]
Soil Association: Fh[5], Mw[2], Fb[2]
Soil Type: SR[7], SWp[2], SWm[2]

PLANT COMMUNITY TYPES (n)

I3.1   sedge fen (15)
I3.2   marsh reed grass fen (5)

I3 graminoid rich fen - Plant community types and approximate cover classes

	13.1	13.2	
SHRUB			
willow	■	■	*Salix* spp. *
FORB			
buck-bean	▮		*Menyanthes trifoliata*
wild mint		■	*Mentha arvensis*
marsh skullcap	■		*Scutellaria galericulata*
GRASS			
sedge	▮		*Carex* spp.
marsh reed grass		▮	*Calamagrostis canadensis*
narrow reed grass		▮	*Calamagrostis stricta*
MOSS			
brown moss	▮		*Drepanocladus* spp.

Approximate cover classes (%): ■ <2   ▪ 2–4.99   ▮ 5–9.99   ▬ 10–19.99   ▬ ≥20.
* Characteristic species with a prominence value <20 for all plant community types.

GENERAL DESCRIPTION

The marsh ecosite is found in level and depressional areas and around the shorelines of water bodies and riparian zones. The water is above the rooting zone for at least a portion of the growing season. This ecosite is dominated by a diversity of emergent sedges and rushes.

d2 low-bush cranberry tA　　　　**m1 marsh**

SUCCESSIONAL RELATIONSHIPS

The marsh ecosite is near the beginning stages of hydrarch succession. The marsh ecosite can be thought of as successionally stable with changes in plant community composition being determined largely by disturbance regime and water level.

MANAGEMENT INTERPRETATIONS

Drought Limitations	Excess Moisture	Rutting Hazard	Compaction Hazard	Puddling Hazard	Erosion Hazard	Frost Heave Hazard	Soil Temperature Limitations	Vegetation Competition	Windthrow Hazard	Productivity	Season of Harvest
L	H	H	L	L	H			not applicable			

hydric/rich

Nutrient Regime

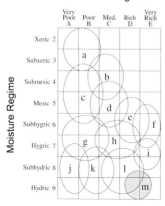

INDICATOR SPECIES

cattail
reed grass
sedge
rush
bulrush

PRODUCTIVITY

not applicable

SITE CHARACTERISTICS

Moisture Regime: hydric[6], subhydric[3]
Nutrient Regime: rich[8], medium[1]
Topographic Position: level[7], depression[3]

SOIL CHARACTERISTICS

Organic Thickness: $(0–5)^3$, $≥80^3$, $(6–15)^3$
Humus Form: peatymor[4], none[4]
Surface Texture: fibric[3], mS[1], cS[1]
Effective Texture: SiS[2], mS[1], cS[1], mesic[1], fibric[1]
Depth to Mottles/Gley: $(0–25)^5$
Drainage: very poor[7], poor[3]
Depth to Water Table: $(0–25)^8$, $(26–50)^1$
Parent Material: F[4], L[2], O[1], H[1]
Soil Subgroup: R.G[6], TY.F[1]
Soil Association: Mh[6], Fs[1], Av[1]
Soil Type: SWm[7], SR[3]

ECOSITE PHASES (n)

m1 marsh (15)

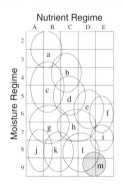

Nutrient Regime

Moisture Regime

CHARACTERISTIC SPECIES

Shrub

[3] willow

Forb

[12] cattail
[6] northern willowherb
[3] wild mint

Grass

[16] reed grass
[12] sedge

Moss

[10] brown moss

SITE CHARACTERISTICS

Moisture Regime: hydric[6], subhydric[3]
Nutrient Regime: rich[8], medium[1]
Topographic Position: level[7], depression[3]

SOIL CHARACTERISTICS

Organic Thickness: (0–5)[3], ≥80[3], (6–15)[3]
Humus Form: peatymor[4], none[4]
Surface Texture: fibric[3], mS[1], cS[1]
Effective Texture: SiS[2], mS[1], cS[1], mesic[1], fibric[1]
Depth to Mottles/Gley: (0–25)[5]
Drainage: very poor[7], poor[3]
Depth to Water Table: (0–25)[8], (26–50)[1]
Parent Material: F[4], L[2], O[1], H[1]
Soil Subgroup: R.G[6], TY.F[1]
Soil Association: Mh[6], Fs[1], Av[1]
Soil Type: SWm[7], SR[3]

PLANT COMMUNITY TYPES (n)

m1.1 cattail marsh (6)
m1.2 reed grass marsh (6)
m1.3 bulrush marsh (3)

m1 marsh - Plant community types and approximate cover classes

		m1.1	m1.2	m1.3	
SHRUB					
	willow	▮			*Salix* spp.
FORB					
	cattail	▮	▮		*Typha latifolia*
	northern willowherb	▮			*Epilobium ciliatum*
	wild mint	▮			*Mentha arvensis*
GRASS					
	reed grass	▮	▮	▮	*Phalaris arundinacea*/*Phragmites australis*
	sedge	▮	▮		*Carex* spp.
	bulrush			▮	*Scirpus* spp.
	marsh reed grass	▮ ·	▮		*Calamagrostis canadensis*
	creeping spike-rush	· ▮			*Eleocharis palustris**
	rush	▮	·		*Juncus* spp.*
MOSS					
	brown moss	▮	▮		*Drepanocladus* spp.

Approximate cover classes (%): ▪ <2   ▬ 2–4.99   ▬ 5–9.99   ▬ 10–19.99   ▬ ≥20.
* Characteristic species with a prominence value <20 for all plant community types.

9.0 SOIL TYPE CLASSIFICATION

Soil types are taxonomic units used to stratify soils based on moisture regime, effective soil texture, organic matter thickness, and solum depth. Soil types can be used independently, in association with the hierarchical classification system (ecosite, ecosite phase, and plant community type) or to classify disturbed sites.

Along with moisture regime, organic matter thickness, and solum depth, effective texture is central to the soil type classification system. Effective texture for mineral soils is generally defined as the textural class of the finest-textured horizon that occurs 20 to 60 cm below the mineral soil surface and that is at least 10 cm thick. The 10-cm minimum thickness stipulation avoids misclassifying fine-textured soils that are predominantly coarse, but have thin, finer-textured depositional bands.

Figure 29 illustrates a soil profile in which two sandy loam layers overlay two distinct layers of sandy clay and loam. Sandy clay is the soil's effective texture because it is the finest textural horizon in the 20 to 60 cm zone and its thickness meets or exceeds 10 cm.

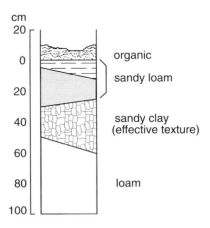

Figure 29.   An example of a soil profile illustrating effective texture.

The effective texture for organic soils is the dominant organic matter decomposition class (fibric, mesic, or humic), 40 to 80 cm below the surface (see Appendix 1).

A soil type is represented by a two- or three-character code (Figure 30). When used with the hierarchical system the soil code is separated from the ecological unit (ecosite, ecosite phase, or plant community type) with a slash (/). The first letter in the soil type code is an S (soil type identifier) followed by a capital V, D, M, W, R, or S that represents very dry, dry, moist, wet, organic, and shallow soils, respectively (Figure 30, Table 4). The moisture regime classes that define very dry (V), dry (D), moist (M), and wet (W), and the textural classes and conditions that define each soil type are presented in Table 4.

The third character of the soil type code, when required, is a number that represents the texture in the very dry, dry, and moist soil types (Figure 30) or a letter that represents peaty (p) with the moist and wet soil types, or mineral (m) with the wet soil types (Table 4). A graphic key to the soil types is provided in Figure 31.

The Organic soil type (SR) represents soils that meet the requirements of the Organic soil order in *The Canadian system of soil classification* (Agriculture Canada Expert Committee on Soil Survey 1987). An organic soil must have an accumulation of organic matter at least 60 cm thick if the material is fibric or at least 40 cm thick if the material is dominantly mesic or humic.

The Shallow soil type (SS) has bedrock within 30 cm of the mineral soil surface.

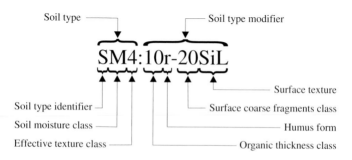

Figure 30.   An example of a typical soil type naming convention.

Table 4.   Soil type description

Soil type		Effective texture/ soil conditions
Very dry soils		**Very xeric-xeric-subxeric**
SV1	Very Dry/Sandy	sand, loamy sand
SV2	Very Dry/Coarse Loamy	sandy loam, silty sand
SV3	Very Dry/Silty-Loamy	silt, silt loam, loam
SV4	Very Dry/Fine Loamy-Clayey	silty clay loam, clay loam, sandy clay loam, sandy clay, silty clay, clay, heavy clay
Dry soils		**Submesic**
SD1	Dry/Sandy	sand, loamy sand
SD2	Dry/Coarse Loamy	sandy loam, silty sand
SD3	Dry/Silty-Loamy	silt, silt loam, loam
SD4	Dry/Fine Loamy-Clayey	silty clay loam, clay loam, sandy clay loam, sandy clay, silty clay, clay, heavy clay
Moist soils		**Mesic-subhygric**
SM1	Moist/Sandy	sand, loamy sand
SM2	Moist/Coarse Loamy	sandy loam, silty sand
SM3	Moist/Silty-Loamy	silt, silt loam, loam
SM4	Moist/Fine Loamy-Clayey	silty clay loam, clay loam, sandy clay loam, sandy clay, silty clay, clay, heavy clay
SMp	Moist/Peaty	any soil texture; organic thickness 20 cm or greater
Wet soils		**Hygric-subhydric-hydric**
SWm	Wet/Mineral	any soil texture; organic thickness less than 20 cm
SWp	Wet/Peaty	any soil texture; organic thickness 20 cm or greater
Organic soils		**Hygric-subhydric-hydric**
SR	Organic	belonging to the Organic soil order
Shallow soils		**Any moisture regime**
SS	Shallow	any soil texture; bedrock encountered within 30 cm of the mineral soil surface

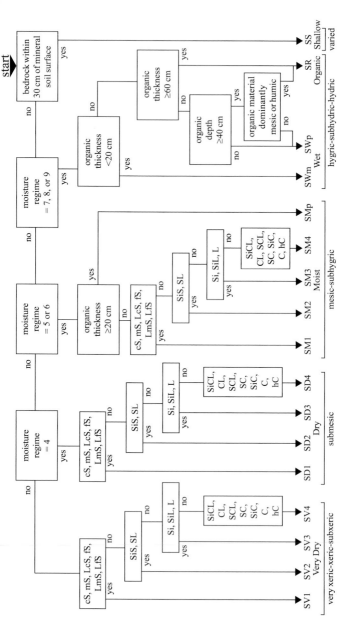

Figure 31. Graphic key to the soil types.

Soil attributes were used as soil type modifiers if they significantly influenced tree productivity (based on site index) or were considered to have important implications toward resource management. Soil type modifiers include organic thickness, humus form, surface coarse fragments, and surface texture (Table 5, Figure 30).

Soil type modifier codes and classes are outlined in Table 5 and a brief description of each is provided below.

Organic thickness

Six organic thickness classes have been defined with a code for each class approximating the mean organic thickness value for the class (Table 5).

Humus form

Five classes of humus form have been identified and outlined in Table 5. Several tools to assist in the classification of humus forms are provided in Section 5.0.

Surface coarse fragments

Surface coarse fragments are the percentage of particles 2 mm or greater in the upper 20 cm of mineral soil. Six classes have been defined with a code for each class approximating the mean percent value for the class (Table 5).

Surface texture

Mineral soils

Surface texture for mineral soils is defined as the dominant texture within the top 20 cm. If the top 10 cm of soil is a uniform texture, use it as the surface texture. If, for example, the uppermost soil horizon is silt loam in texture and is only 5 cm thick while the horizon below it is clay in texture, then the clay layer is considered the surface texture. Textural classes such as "VC" representing very coarse, or naming conventions for common textures such as "SiL" can be used as the code for the surface texture soil modifier.

Organic soils

Surface texture for organic soils is defined as the dominant decomposition class within the top 40 cm of organic material. Three classes have been identified: fibric, mesic, and humic, and are based on the von Post scale of decomposition (Appendix 1).

Table 5.   Soil type modifier codes and classes

Description	Code	Class
Organic	3	(0–5)
thickness	8	(6–10)
(cm)	20	(16–25)
	35	(26–39)
	50	(40–59)
	80	≥60
Humus form	r	mor
	w	raw moder
	d	moder
	u	mull
	p	peatymor
Surface coarse	2	(0–4)
fragments (%)[a]	10	(5–14)
	20	(15–29)
	40	(30–49)
	60	(50–79)
	90	≥80
Surface texture[b]	VC	very coarse
	cS	coarse sand
	mS	medium sand
	LcS	loamy coarse sand
	C	coarse
	fS	fine sand
	LmS	loamy medium sand
	LfS	loamy fine sand

[a] Total percent coarse fragments >2 mm in diameter.
[b] Surface texture is defined as the dominant soil texture within the upper 20 cm for mineral soils and is the dominant decomposition class of organic material within the upper 40 cm for organic soils.

Table 5.   concluded

Description	Code	Class
Surface texture continued	MC	moderately coarse
	SiS	silty sand
	SL	sandy loam
	M	medium
	Si	silt
	SiL	silt loam
	L	loam
	MF	moderately fine
	SCL	sandy clay loam
	CL	clay loam
	SiCL	silty clay loam
	F	fine
	SC	sandy clay
	C	clay
	SiC	silty clay
	hC	heavy clay
	R	organic
	Fi	fibric
	Me	mesic
	Hu	humic

GENERAL DESCRIPTION

Very dry, sandy, and loamy sand soils that commonly develop in glaciofluvial and eolian parent materials.

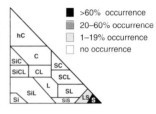

- ■ >60% occurrence
- ▨ 20–60% occurrence
- ░ 1–19% occurrence
- □ no occurrence

SOIL CHARACTERISTICS

Organic Thickness: $(0–5)^8$, $(6–15)^3$
Humus Form: mor^7, raw $moder^2$
Surface Texture: mS^5, cS^2, fS^1, LmS^1
Surface Coarse Fragments: $(0–4)^7$, $(5–14)^1$, $(15–29)^1$
Effective Texture: mS^5, cS^2, LmS^1
Depth to Mottles/Gley: $none^{10}$
Drainage: $rapid^7$, very $rapid^2$, $well^1$
Depth to Carbonates: $none^{10}$
Parent Material: GF^5, FE^1, E^1
Soil Subgroup: $E.EB^6$, $O.EB^1$, $E.DYB^1$
Soil Association: Pn^5, Kk^3, Bd^1, Bx^1

SITE CHARACTERISTICS

Moisture Regime: $subxeric^6$, $xeric^3$, very $xeric^1$
Nutrient Regime: $poor^9$
Topographic Position: $midslope^3$, $level^3$, $crest^2$, upper $slope^1$

ASSOCIATED ECOSITES

Ecosite: a^5, b^4

COMMENTS

These soils are closely associated with the Pine soil association (Pn), and the lichen ecosite (a). SV1 soils are nutrient poor and have a low capacity to retain water because they have a deep coarse-textured solum. Forest productivity on these soils tends to be low. A moderate windthrow hazard exists for shallow-rooted white spruce trees.

MANAGEMENT INTERPRETATIONS

Drought Limitations	Excess Moisture	Rutting Hazard	Compaction Hazard	Puddling Hazard	Erosion Hazard	Frost Heave Hazard	Soil Temperature Limitations	Vegetation Competition	Windthrow Hazard	Productivity	Season of Harvest
H	L	L	L	L	L	L	L	L	L-M	L	W-Sp

Very Dry/Coarse Loamy (n = 4)

GENERAL DESCRIPTION

Very dry, coarse loamy soils that commonly develop in glaciofluvial and fluviolacustrine parent materials.

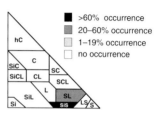

- ■ >60% occurrence
- ▨ 20–60% occurrence
- ▨ 1–19% occurrence
- □ no occurrence

SOIL CHARACTERISTICS

Organic Thickness: (0–5)[10]
Humus Form: mor[10]
Surface Texture: SiS[8], Si[3]
Surface Coarse Fragments: (0–4)[10]
Effective Texture: SiS[8], SL[3]
Depth to Mottles/Gley: none[10]
Drainage: rapid[5], well[5]
Depth to Carbonates: none[10]
Parent Material: GF[5], GF/GL[3], FL[3]
Soil Subgroup: BR.GL[7], E.EB[3]
Soil Association: Wt[3], Pn[3], Ft[3]

SITE CHARACTERISTICS

Moisture Regime: subxeric[8], xeric[3]
Nutrient Regime: poor[10]
Topographic Position: level[5], upper slope[3], crest[3]

ASSOCIATED ECOSITES

Ecosite: a[3], b[3], c[3]

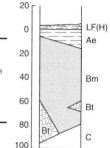

COMMENTS

The SV2 soil type occurs on slope crests to level topography and is strongly influenced by its rapid to well drained, moderately coarse-textured profile. Droughty conditions persist for a significant portion of the growing season.

MANAGEMENT INTERPRETATIONS

Drought Limitations	Excess Moisture	Rutting Hazard	Compaction Hazard	Puddling Hazard	Erosion Hazard	Frost Heave Hazard	Soil Temperature Limitations	Vegetation Competition	Windthrow Hazard	Productivity	Season of Harvest
H	L	L	L	L	L-M	L	L	L	L	L	W-Sp

GENERAL DESCRIPTION

Very dry, silty to loamy soils.

COMMENTS

SV3 soils were not sampled in the Mid-Boreal ecoregions of Saskatchewan but are expected to occur in topographic positions that shed water such as slope crests and steep, south-facing valley slopes where solar radiation is intense. Those soils that do occur on steep slopes are highly susceptible to water erosion.

MANAGEMENT INTERPRETATIONS

Drought Limitations	Excess Moisture	Rutting Hazard	Compaction Hazard	Puddling Hazard	Erosion Hazard	Frost Heave Hazard	Soil Temperature Limitations	Vegetation Competition	Windthrow Hazard	Productivity	Season of Harvest
H	L	L	L-M	M	H	L-M	L	L	L	L	A

Very Dry/Fine Loamy-Clayey (n=0)

GENERAL DESCRIPTION

Very dry, fine loamy to clayey soils.

COMMENTS

SV4 soils were not sampled in the Mid-Boreal ecoregions of Saskatchewan but are expected to occur in topographic positions that shed water such as slope crests and steep, south-facing valley slopes where solar radiation is intense. Those soils that do occur on steep slopes are highly susceptible to water erosion.

MANAGEMENT INTERPRETATIONS

Drought Limitations	Excess Moisture	Rutting Hazard	Compaction Hazard	Puddling Hazard	Erosion Hazard	Frost Heave Hazard	Soil Temperature Limitations	Vegetation Competition	Windthrow Hazard	Productivity	Season of Harvest
H	L	L	L-M	M	H	L-M	L	L	L	L	A

SOIL TYPE

SD1 — Dry/Sandy (n = 37)

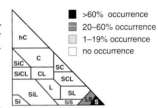

>60% occurrence
20–60% occurrence
1–19% occurrence
no occurrence

GENERAL DESCRIPTION

Dry, sandy soils that most commonly develop in glaciofluvial and fluvio-lacustrine parent materials.

SOIL CHARACTERISTICS

Organic Thickness: $(6–15)^6$, $(0–5)^4$
Humus Form: mor^6, raw $moder^4$
Surface Texture: mS^4, LmS^3, fS^1, LfS^1
Surface Coarse Fragments: $(0–4)^8$, $(5–14)^1$
Effective Texture: mS^5, LmS^2, fS^1, cS^1, LfS^1
Depth to Mottles/Gley: $none^9$, $(51-100)^1$
Drainage: $well^6$, $rapid^4$
Depth to Carbonates: $none^{10}$
Parent Material: GF^4, FL^2, F^1, GF/M^1
Soil Subgroup: $E.EB^4$, $BR.GL^3$, $E.DYB^2$
Soil Association: Wt^4, Pn^3

SITE CHARACTERISTICS

Moisture Regime: $submesic^{10}$
Nutrient Regime: $poor^6$, $medium^4$
Topographic Position: $midslope^3$, upper $slope^3$, $level^1$, lower $slope^1$

ASSOCIATED ECOSITES

Ecosite: d^4, b^3, c^2

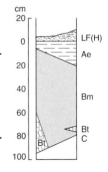

COMMENTS

SD1 soils exhibit rapid to well internal soil drainage. Typically, the upper 60 cm of the soil profile is uniformly coarse-textured although occasionally, thin, sandy loam to clayey fluviolacustrine bands (1–3 cm) may be present. These bands or other finer-textured parent materials may also be present at depths greater than 60 cm, which helps reduce the potential for severe droughty soil conditions.

MANAGEMENT INTERPRETATIONS

Drought Limitations	Excess Moisture	Rutting Hazard	Compaction Hazard	Puddling Hazard	Erosion Hazard	Frost Heave Hazard	Soil Temperature Limitations	Vegetation Competition	Windthrow Hazard	Productivity	Season of Harvest
M	L	L	L	L	L	L	L	M	L-M	M	A

Dry/Coarse Loamy (n = 13) | SD2

GENERAL DESCRIPTION

Dry, coarse loamy soils that develop in a variety of parent materials.

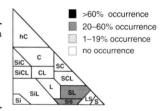

- ■ >60% occurrence
- ▨ 20–60% occurrence
- ░ 1–19% occurrence
- □ no occurrence

SOIL CHARACTERISTICS

Organic Thickness: (6–15)[8], (0–5)[2]
Humus Form: mor[9]
Surface Texture: SiS[5], SL[2]
Surface Coarse Fragments: (0–4)[7], (5–14)[2], (50–79)[1], (15–29)[1]
Effective Texture: SL[5], SiS[5]
Depth to Mottles/Gley: none[9]
Drainage: well[8], rapid[2]
Depth to Carbonates: none[7], >50[3]
Parent Material: GF/M[2], GF[2], M[2], GL[2], FL[2]
Soil Subgroup: BR.GL[7], E.EB[2]
Soil Association: Ln[3], Bt[3], Wt[2], Pn[1], Ft[1], Lc[1], Kk[1]

SITE CHARACTERISTICS

Moisture Regime: submesic[10]
Nutrient Regime: poor[5], medium[5]
Topographic Position: midslope[5], lower slope[3], level[2]

ASSOCIATED ECOSITES

Ecosite: d[5], c[2], b[2]

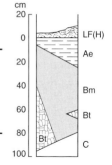

COMMENTS

SD2 soils occur on midslope to level topographic positions. This soil type is associated with similar environmental site characteristics as SD1. Droughty soil conditions may be moderated by the presence of thin, fluviolacustrine bands or finer-textured depositions below 60 cm. Mottles are typically not encountered in the soil profile.

MANAGEMENT INTERPRETATIONS

Drought Limitations	Excess Moisture	Rutting Hazard	Compaction Hazard	Pudding Hazard	Erosion Hazard	Frost Heave Hazard	Soil Temperature Limitations	Vegetation Competition	Windthrow Hazard	Productivity	Season of Harvest
M	L	L	L	L	L-M	L	L	M	L	M	A

SD3 | Dry/Silty-Loamy (n = 4)

GENERAL DESCRIPTION

Dry, silty to loamy soils that commonly develop in glaciolacustrine and fluvio-lacustrine parent materials.

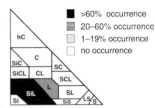

- ■ >60% occurrence
- ▨ 20–60% occurrence
- ▢ 1–19% occurrence
- □ no occurrence

SOIL CHARACTERISTICS

Organic Thickness: $(6–15)^5$, $(0–5)^5$
Humus Form: mor[8], raw moder[3]
Surface Texture: SiL[5], Si[3], LmS[3]
Surface Coarse Fragments: $(0–4)^8$, $(5–14)^3$
Effective Texture: SiL[8], L[3]
Depth to Mottles/Gley: none[10]
Drainage: well[10]
Depth to Carbonates: none[5], >50[5]
Parent Material: GL[5], FL/M[3], FL[3]
Soil Subgroup: BR.GL[10]
Soil Association: Wt[5], Lc[3], Do[3]

SITE CHARACTERISTICS

Moisture Regime: submesic[10]
Nutrient Regime: medium[10]
Topographic Position: midslope[5], level[3], upper slope[3]

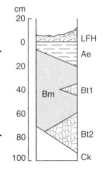

ASSOCIATED ECOSITES

Ecosite: d[5], e[3], b[3]

COMMENTS

SD3 soils occur in upper slope to level positions in the landscape and are most strongly associated with glaciolacustrine deposits. Varves may be evident in the soil profile characteristic of the La Corne (Lc) and the Dorintosh (Do) soil associations. Mottles are not typically encountered in SD3 soils as they tend to be well drained.

MANAGEMENT INTERPRETATIONS

Drought Limitations	Excess Moisture	Rutting Hazard	Compaction Hazard	Puddling Hazard	Erosion Hazard	Frost Heave Hazard	Soil Temperature Limitations	Vegetation Competition	Windthrow Hazard	Productivity	Season of Harvest
M	L	L-M	M	M	H	M	L	M	L	M	A

Dry/Fine Loamy-Clayey (n = 50) SD4

GENERAL DESCRIPTION

Dry, clayey soils that most commonly develop in morainal and glaciofluvial over morainal parent materials.

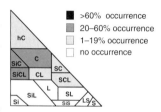

■ >60% occurrence
▨ 20–60% occurrence
□ 1–19% occurrence
□ no occurrence

SOIL CHARACTERISTICS

Organic Thickness: (6–15)[6], (0–5)[4]
Humus Form: mor[9], raw moder[1]
Surface Texture: SL[2], SiL[2], SiS[1], LmS[1], Si[1], LcS[1]
Surface Coarse Fragments: (0–4)[5], (5–14)[3], (15–29)[2]
Effective Texture: SiC[2], SC[2], C[2], SCL[1], CL[1]
Depth to Mottles/Gley: none[10]
Drainage: well[6], mod. well[3], rapid[1]
Depth to Carbonates: none[6], >50[3]
Parent Material: M[6], GF/M[1]
Soil Subgroup: BR.GL[7], O.GL[3]
Soil Association: Ln[4], Bt[3], Lc[1], Hw[1]

SITE CHARACTERISTICS

Moisture Regime: submesic[10]
Nutrient Regime: medium[7], poor[2]
Topographic Position: midslope[4], upper slope[3], level[2], crest[1]

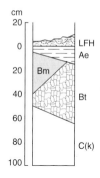

ASSOCIATED ECOSITES

Ecosite: d[6], b[3], c[1]

COMMENTS

SD4 soils occur in upland landscape positions and are characterized by moderately coarse to medium-textured surface layers overlying a fine-textured Bt horizon. This illuviated horizon (Bt) can become extremely hard after extended periods of warm, dry weather. Under these conditions, root development and plant growth are inhibited.

MANAGEMENT INTERPRETATIONS

Drought Limitations	Excess Moisture	Rutting Hazard	Compaction Hazard	Puddling Hazard	Erosion Hazard	Frost Heave Hazard	Soil Temperature Limitations	Vegetation Competition	Windthrow Hazard	Productivity	Season of Harvest
M	L	M	M	H	H	M	L	M	L	M	A

GENERAL DESCRIPTION

Moist, sandy soils that most commonly develop in glaciofluvial and glaciofluvial over morainal parent materials.

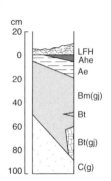

- ■ >60% occurrence
- ▨ 20–60% occurrence
- ▢ 1–19% occurrence
- □ no occurrence

SOIL CHARACTERISTICS

Organic Thickness: $(6–15)^8$, $(0–5)^2$
Humus Form: raw moder[5], mor[4], moder[1]
Surface Texture: mS^3, LmS^3
Surface Coarse Fragments: $(0–4)^8$, $(5–14)^2$
Effective Texture: LmS^4, mS^3, LcS^1
Depth to Mottles/Gley: none[6], $(0–25)^4$
Drainage: imperfect[3], well[3], mod. well[3], rapid[1]
Depth to Carbonates: none[8], >50[1]
Parent Material: GF[4], L[2], GF/M[1], F[1]
Soil Subgroup: E.EB[3], GLE.DYB[1], E.DYB[1], BR.GL[1]
Soil Association: Bt[3], Bx[2], Ln[1], Kk[1], Hw[1], Ft[1], Av[1]

SITE CHARACTERISTICS

Moisture Regime: mesic[5], subhygric[5]
Nutrient Regime: medium[6], poor[3], rich[1]
Topographic Position: midslope[3], upper slope[2], level[2], lower slope[2]

ASSOCIATED ECOSITES

Ecosite: d[6], e[2], g[1], b[1]

COMMENTS

Two typical soil profiles are associated with SM1 soils. One variation exhibits a continuous coarse-textured profile and occurs in the lower topographic position where seepage may occur. The other variation is characterized by a deep coarse-textured layer overlying finer-textured material occurring below 60 cm. Faint mottles may be present in any horizon while distinct to prominent mottles may occur 50 cm below the mineral soil surface.

MANAGEMENT INTERPRETATIONS

Drought Limitations	Excess Moisture	Rutting Hazard	Compaction Hazard	Puddling Hazard	Erosion Hazard	Frost Heave Hazard	Soil Temperature Limitations	Vegetation Competition	Windthrow Hazard	Productivity	Season of Harvest
L	L-M	L	L	L	L	L	L-M	M-H	L-M	H	A

Moist/Coarse Loamy (n = 9) $\boxed{\textbf{SM2}}$

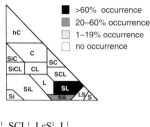

■ >60% occurrence
▨ 20–60% occurrence
▢ 1–19% occurrence
□ no occurrence

GENERAL DESCRIPTION

Moist, coarse loamy soils that develop in a variety of parent materials.

SOIL CHARACTERISTICS

Organic Thickness: (0–5)[8], (6–15)[1], (16–25)[1]
Humus Form: raw moder[4], mor[4], moder[1]
Surface Texture: SL[2], SiS[1], SiL[1], SiC[1], Si[1], SCL[1], LcS[1], L[1]
Surface Coarse Fragments: (0–4)[8], (5–14)[1], (30–49)[1]
Effective Texture: SL[7], SiS[3]
Depth to Mottles/Gley: (0–25)[6], none[3], (26–50)[1]
Drainage: well[3], mod. well[3], imperfect[3]
Depth to Carbonates: none[6], (0–29)[4]
Parent Material: GL[2], GF/M[2], L[1], GF/GL[1], GF[1], FE[1], F/GL[1]
Soil Subgroup: BR.GL[4], O.LG[3], O.HG[1], GLBR.GL[1], GL.GL[1]
Soil Association: Lc[4], Bt[4], Ft[2]

SITE CHARACTERISTICS

Moisture Regime: mesic[6], subhygric[4]
Nutrient Regime: medium[4], poor[3], rich[2]
Topographic Position: lower slope[4], midslope[2], level[2], toe[1]

ASSOCIATED ECOSITES

Ecosite: d[6], g[2], e[1], c[1]

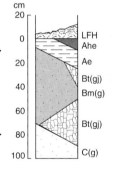

COMMENTS

Two typical soil profiles are associated with SM2. Those with a continuous, moderately coarse-textured profile are strongly influenced by its location in lower slope positions where water is discharged from upslope sources. Those SM2 soils with finer-textured discontinuous layers at depths 60 cm or greater are strongly influenced by the impeded drainage caused by this underlying layer.

MANAGEMENT INTERPRETATIONS

Drought Limitations	Excess Moisture	Rutting Hazard	Compaction Hazard	Puddling Hazard	Erosion Hazard	Frost Heave Hazard	Soil Temperature Limitations	Vegetation Competition	Windthrow Hazard	Productivity	Season of Harvest
L	L-M	L-M	L-M	L	M	L-M	L-M	M-H	L	H	A

SOIL TYPE

GENERAL DESCRIPTION

Moist, silty to loamy soils that develop in fluvial, glaciolacustrine, and fluvio-lacustrine parent materials.

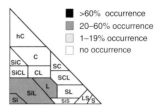

■ >60% occurrence
▨ 20–60% occurrence
□ 1–19% occurrence
□ no occurrence

SOIL CHARACTERISTICS

Organic Thickness: $(6–15)^7$, $(0–5)^2$
Humus Form: mor[6], raw moder[2], moder[2]
Surface Texture: Si[4], SiL[3], SL[1]
Surface Coarse Fragments: $(0–4)^8$
Effective Texture: SiL[4], Si[3], L[3]
Depth to Mottles/Gley: $(0–25)^5$, none[4]
Drainage: mod. well[4], imperfect[4], well[2]
Depth to Carbonates: none[7], $(0–29)^2$
Parent Material: F[5], GL[2], FL[1]
Soil Subgroup: BR.GL[2], R.G[1], GL.HR[1], O.G[1]
Soil Association: Wt[3], Sk[2], Lc[2], Bt[1], Av[1], Aw[1]

SITE CHARACTERISTICS

Moisture Regime: subhygric[7], mesic[3]
Nutrient Regime: medium[5], rich[4]
Topographic Position: level[4], midslope[2], toe[2], lower slope[1]

ASSOCIATED ECOSITES

Ecosite: d[5], h[2], f[2], e[1]

```
cm
20
                    LFH
 0                  Ahe
          Bm(g)
20                  Ahb
                    Bt
40

60
                    Bt(g)
80
100                 C(g)
```

COMMENTS

SM3 soils typically occur on level, fluvially deposited landscapes. Soils in this environment may exhibit buried, humified Ah horizons (Ahb). High hazard ratings generally apply to those SM3 soils that are associated with sites that have a subhygric moisture regime. Mottling may be present in any horizon.

MANAGEMENT INTERPRETATIONS

Drought Limitations	Excess Moisture	Rutting Hazard	Compaction Hazard	Puddling Hazard	Erosion Hazard	Frost Heave Hazard	Soil Temperature Limitations	Vegetation Competition	Windthrow Hazard	Productivity	Season of Harvest
L	L-M	M-H	M-H	M-H	M-H	M-H	L-M	H	L	H	A

Moist/Fine Loamy-Clayey (n = 233) SM4

GENERAL DESCRIPTION

Moist, fine loamy to clayey soils that most commonly develop in morainal and glaciolacustrine parent materials.

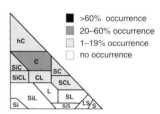

■ >60% occurrence
▨ 20–60% occurrence
▢ 1–19% occurrence
▢ no occurrence

SOIL CHARACTERISTICS

Organic Thickness: $(6-15)^8$, $(0-5)^2$
Humus Form: mor[6], raw moder[4]
Surface Texture: SiL[2], SL[2], Si[1], L[1], LmS[1]
Surface Coarse Fragments: $(0-4)^7$, $(5-14)^2$
Effective Texture: C[4], SiC[2], hC[1], SCL[1], SiCL[1], CL[1]
Depth to Mottles/Gley: none[7], $(0-25)^2$
Drainage: mod. well[6], well[2], imperfect[2]
Depth to Carbonates: none[5], >50[3], $(30-50)^1$
Parent Material: M[6], GL[1]
Soil Subgroup: O.GL[3], BR.GL[3]
Soil Association: Ln[3], Bt[2], Do[2], Wv[1]

SITE CHARACTERISTICS

Moisture Regime: mesic[8], subhygric[2]
Nutrient Regime: medium[7], rich[2]
Topographic Position: level[3], midslope[3], upper slope[2], lower slope[1]

ASSOCIATED ECOSITES

Ecosite: d[7], e[1]

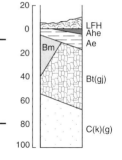

COMMENTS

SM4 was the most extensively sampled soil type in the study area and occurs in upland landscape positions. Typically, these soils have a medium to moderately coarse-textured surface layer overlying a fine-textured Bt horizon. This illuvial horizon (Bt) may impede internal soil drainage during high rainfall and spring runoff periods causing saturated soil conditions in the upper horizons. SM4 is most strongly associated with the reference low-bush cranberry ecosite (d).

MANAGEMENT INTERPRETATIONS

Drought Limitations	Excess Moisture	Rutting Hazard	Compaction Hazard	Puddling Hazard	Erosion Hazard	Frost Heave Hazard	Soil Temperature Limitations	Vegetation Competition	Windthrow Hazard	Productivity	Season of Harvest
L	L-M	M-H	H	H	M-H	M-H	L-M	H	L	H	A

Moist/Peaty (n = 10)

GENERAL DESCRIPTION

Moist soils with an organic layer thickness equal to or greater than 20 cm that have developed in a variety of parent materials.

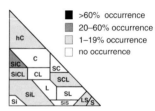

■ >60% occurrence
▨ 20–60% occurrence
▧ 1–19% occurrence
☐ no occurrence

SOIL CHARACTERISTICS

Organic Thickness: (16–25)[6], (26–39)[4]
Humus Form: mor[8], raw moder[2]
Surface Texture: variable
Surface Coarse Fragments: (0–4)[7], (50–79)[1], (5–14)[1], (30–49)[1]
Effective Texture: variable
Depth to Mottles/Gley: (0–25)[5], none[3], (51–100)[1], (26–50)[1]
Drainage: imperfect[3], poor[3], well[2], mod. well[2]
Depth to Carbonates: none[8], (0–29)[2]
Parent Material: M[2], GF/M[2], GF[2], L[1], GL[1], GF/GL[1], F[1]
Soil Subgroup: R.HG[2], R.G[1], O.LG[1], O.GL[1], O.EB[1], GL.GL[1], GL.EB[1], CU.R[1], BR.GL[1]
Soil Association: Kw[3], Kk[3], Ft[3], Bo[3]

SITE CHARACTERISTICS

Moisture Regime: subhygric[7], mesic[3]
Nutrient Regime: medium[5], rich[3], poor[2]
Topographic Position: lower slope[5], level[3], midslope[1], depression[1]

ASSOCIATED ECOSITES

Ecosite: d[3], e[2], g[2], h[2], l[1]

COMMENTS

The mean moisture regime for SMp is higher than other moist soil types. As compared to a thin organic layer, the thick organic layer of SMp has the inherent quality of reducing potential soil degradation effects of heavy equipment. Faint to distinct mottles commonly occur throughout the soil profile.

MANAGEMENT INTERPRETATIONS

Drought Limitations	Excess Moisture	Rutting Hazard	Compaction Hazard	Puddling Hazard	Erosion Hazard	Frost Heave Hazard	Soil Temperature Limitations	Vegetation Competition	Windthrow Hazard	Productivity	Season of Harvest
L	M	H	H	H	H	H	M-H	H	M-H	H	W

Wet/Mineral (n = 50)

GENERAL DESCRIPTION

Wet soils with an organic layer thickness less than 20 cm.

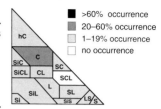

■ >60% occurrence
▨ 20–60% occurrence
░ 1–19% occurrence
□ no occurrence

SOIL CHARACTERISTICS

Organic Thickness: (6–15)[6], (0–5)[3], (16–25)[1]
Humus Form: mor[4], raw moder[2], none[2], mull[1]
Surface Texture: variable
Surface Coarse Fragments: (0–4)[6], (5–14)[3], (15–29)[1]
Effective Texture: variable
Depth to Mottles/Gley: (0–25)[7], none[1], (26–50)[1]
Depth to Water Table: none[4], (0–25)[3], (51–100)[2], (26–50)[1]
Drainage: poor[4], imperfect[3], very poor[2]
Depth to Carbonates: none[6], (0–29)[3], >50[1]
Parent Material: F[3], L[2], M[1], GL[1]
Soil Subgroup: R.G[3], R.HG[2], O.LG[2], O.G[1]
Soil Association: Av[3], Mh[2], Mw[1], Bx[1], Aw[1]

SITE CHARACTERISTICS

Moisture Regime: hygric[6], subhydric[2], hydric[1]
Nutrient Regime: rich[6], poor[2], medium[1]
Topographic Position: level[4], depression[2], lower slope[1], toe[1]

ASSOCIATED ECOSITES

Ecosite: e[3], m[2], g[1], h[1], l[1]

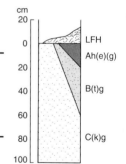

cm
20
0   LFH
20   Ah(e)(g)
40   B(t)g
60
80   C(k)g
100

COMMENTS

SWm sites are associated with forested ecosite phases that are typically situated in seepage zones or on level topography influenced by nutrient-replenishing groundwater sources. The non-forested marsh ecosite (m), situated adjacent to lakes and streams, is also strongly associated with SWm. High operational hazard ratings exist for this soil type based on its high soil water content.

MANAGEMENT INTERPRETATIONS

Drought Limitations	Excess Moisture	Rutting Hazard	Compaction Hazard	Puddling Hazard	Erosion Hazard	Frost Heave Hazard	Soil Temperature Limitations	Vegetation Competition	Windthrow Hazard	Productivity	Season of Harvest
L	H	H	H	H	H	H	H	M-H	H	L-H	W

GENERAL DESCRIPTION

Wet soils with an organic thickness equal to or greater than 20 cm. SWp is commonly associated with ecosites that have feather moss or sphagnum-dominated moss layers.

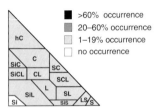

- ■ >60% occurrence
- ▨ 20–60% occurrence
- ☐ 1–19% occurrence
- ☐ no occurrence

SOIL CHARACTERISTICS

Organic Thickness: (26–39)[4], (16–25)[3], (40–59)[3]
Humus Form: peatymor[4], mor[4], moder[1]
Surface Texture: variable
Surface Coarse Fragments: (0–4)[7], (5–14)[1], (15–29)[1]
Effective Texture: variable
Depth to Mottles/Gley: (0–25)[8], none[1], (26–50)[1]
Depth to Water Table: none[4], (26–50)[2], (51–100)[1], >100[1], (0–25)[1]
Drainage: poor[5], very poor[3], imperfect[2]
Depth to Carbonates: none[6], (0–29)[3]
Parent Material: GL[3], M[2], F[2], O/M[1]
Soil Subgroup: R.G[4], O.LG[2], R.HG[1]
Soil Association: Aw[5], Av[2]

SITE CHARACTERISTICS

Moisture Regime: hygric[5], subhydric[4], hydric[1]
Nutrient Regime: rich[4], medium[4], poor[2]
Topographic Position: level[5], depression[2], lower slope[1]

ASSOCIATED ECOSITES

Ecosite: g[2], l[2], h[2], e[1], k[1]

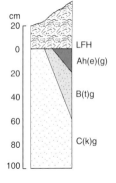

COMMENTS

SWp soils most commonly occur on flat, depressional, or lower slope positions in the landscape where seepage occurs or where local drainage waters accumulate. Most of the tree roots found in SWp occur in its thick peaty layers, increasing the risk of blowdown. Black spruce, tamarack, balsam poplar, and white spruce are the most common tree species associated with SWp. Prominent mottles or gleying are commonly encountered in the soil profile.

MANAGEMENT INTERPRETATIONS

Drought Limitations	Excess Moisture	Rutting Hazard	Compaction Hazard	Puddling Hazard	Erosion Hazard	Frost Heave Hazard	Soil Temperature Limitations	Vegetation Competition	Windthrow Hazard	Productivity	Season of Harvest
L	H	H	H	H	H	H	H	M-H	H	L	W

Organic (n = 115)

SR

GENERAL DESCRIPTION

Organic soils are wet with an organic thickness equal to or greater than 60 cm if the material is fibric, or equal to or greater than 40 cm if the material is mesic or humic. On sites with sphagnum and/or feather mosses covering the surface substrate, microtopography tends to be hummocky.

SOIL CHARACTERISTICS

Organic Thickness: ≥80[9]
Humus Form: peatymor[10]
Surface Texture: fibric[8], mesic[2]
Effective Texture: fibric[5], mesic[3], humic[2]
Drainage: very poor[6], poor[4]
Depth to Water Table: (0–25)[6], none[3], (26–50)[2]
Parent Material: O[4], N[2], O/M[1]
Soil Subgroup: TY.F[3], T.M[2], TY.M[1], T.F[1]
Soil Association: Fh[3], Fs[2], Bf[2], Fb[2], Bb[1]

SITE CHARACTERISTICS

Moisture Regime: subhydric[5], hydric[3], hygric[1]
Nutrient Regime: medium[4], rich[3], poor[2], very poor[1]
Topographic Position: depression[5], level[4]

ASSOCIATED ECOSITES

Ecosite: l[4], j[2], k[2]

cm: 20, 0, 20, 40, 60, 80, 100 — Of, Om, Oh, Cg

COMMENTS

SR soils are typically located on flat or depressional areas in the landscape where regional or local drainage waters accumulate. They exhibit a diverse range of profiles based on organic thickness (40 cm to over 3 m) and on the degree of organic decomposition (poorly to very well decomposed). SR soils are strongly associated with lowland ecosites such as the bog (j), the poor fen (k), and the rich fen (l).

MANAGEMENT INTERPRETATIONS

Drought Limitations	Excess Moisture	Rutting Hazard	Compaction Hazard	Puddling Hazard	Erosion Hazard	Frost Heave Hazard	Soil Temperature Limitations	Vegetation Competition	Windthrow Hazard	Productivity	Season of Harvest
L	H	H	L	L	L	H	H	L	H	L	W

GENERAL DESCRIPTION

Shallow soils with less than or equal to 30 cm of mineral material (any texture) overlying bedrock. This soil type includes exposed bedrock surfaces.

COMMENTS

SS soils were not sampled in the Mid-Boreal ecoregions of Saskatchewan but are expected to occur on the northern edge of the study area in the transition zone bordering the Boreal Shield Ecozone. No interpretations were assessed for SS since moisture regime and texture, two of the most important factors used to assess soil limitations and opportunities, do not define this soil type. If an SS soil type is encountered within a management area, consult Section 12.0, Management Interpretations, for guidelines.

MANAGEMENT INTERPRETATIONS

Drought Limitations	Excess Moisture	Rutting Hazard	Compaction Hazard	Puddling Hazard	Erosion Hazard	Frost Heave Hazard	Soil Temperature Limitations	Vegetation Competition	Windthrow Hazard	Productivity	Season of Harvest

10.0 SELECTED SOIL TYPE PHOTOGRAPHS

Shown are six examples of soil types that are either commonly found within Saskatchewan's Mid-Boreal ecoregions and/or have important characteristics applicable to resource management planning.

◄ SV1 Very dry/Sandy soil

This is an example of an Eluviated Eutric Brunisol of the Pine soil association (Pn) developed in coarse-textured glaciofluvial parent material. This soil is rapidly drained and supports a jP/bearberry/lichen plant community type.

▲ SD2 Dry/Coarse loamy soil

This is an example of a Brunisolic Gray Luvisol of the Waterhen soil association (Wt) developed in reworked, moderately coarse-textured fluvio-lacustrine material. Note the presence of thin, finer-textured bands of material at 45 cm and 65 cm below the mineral soil surface.

SM4   Moist/Fine loamy-clayey soil ▶

This is an example of a Gleyed Gray Luvisol of the Loon soil association (Ln) developed in morainal parent material.

◀ SWm   Wet/Mineral soil

This is an example of an Orthic Gleysol of the Kewanoke soil association (Kk) developed in coarse-textured glaciofluvial parent material. This prominently mottled soil supports a bS-jP/Labrador tea/ feather moss plant community type.

◀ SWp   Wet/Peaty soil

This is an example of an Orthic Luvic Gleysol of the Dorintosh soil association (Do) developed in fine-textured glaciolacustrine parent material. This soil supports a wS-bS/Labrador tea/horsetail plant community type.

SR   Organic soil   ▶

This is an example of a Typic Fibrisol formed in a flat bog. This soil supports a black spruce–Labrador tea/cloudberry/peat moss plant community type.

11.0   PLANT RECOGNITION

This section illustrates 103 plant species that are common in the Mid-Boreal ecoregions of Saskatchewan. The illustrations are provided to assist users in the identification of these plants and to facilitate the use of the field guide. Included with the picture of each plant is an edatopic grid (moisture/nutrient grid) for each species showing the relationship of the species occurrence to moisture and nutrient regimes. The individual grids were developed from actual frequency of occurrence data derived from the plot data. The distribution of a species on the edatope may vary by local area. The most frequent occurrence of a species is shown as the dark shaded area on the grid while the general range of observations is shown as the gray shaded area. Species do occur outside the range given in the diagrams as they do not show the extreme ranges for each species.

The species have been arranged alphabetically by scientific name in six growth form groups: tree, shrub, forb, grass, moss, and lichen. The forb group includes several plants regarded by some authorities as dwarf shrubs, including *Rubus chamaemorus* and *Rubus pubescens*. Nomenclature follows Farrar (1995) for trees, Moss (1983) for other vascular plants, Ireland et al. (1987) and Schofield (1992) for mosses, Stotler and Crandall-Stotler (1977) for liverworts, and Egan (1987) for lichens.

For more information about the plants in this section and other plants in the Mid-Boreal ecoregions of Saskatchewan consult Appendix 8 for a reference list. If a plant is not included in this guide or you require additional information, consult one of the following: Johnson et al. (1995), MacKinnon et al. (1992), Vitt et al. (1988), Moss (1983), Looman and Best (1979), Hitchcock and Cronquist (1973), or Ireland (1982).

An index of scientific and common names for the plants illustrated here appears at the end of the field guide.

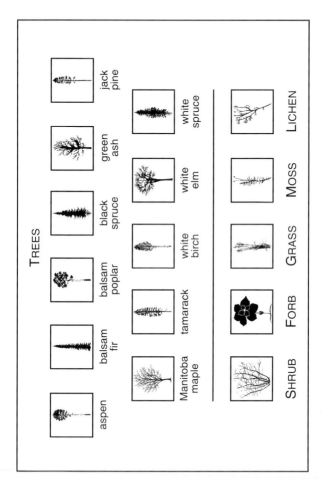

Figure 32. Growth form icons.

Abies balsamea (L.) Mill.
balsam fir

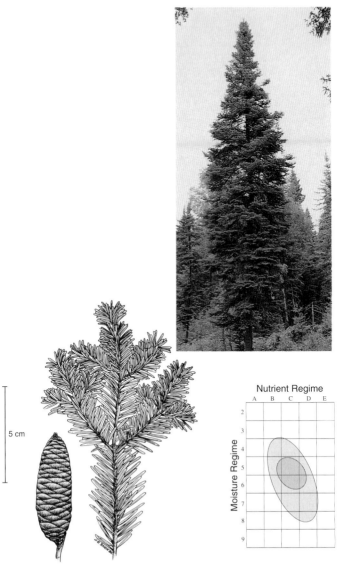

5 cm

Nutrient Regime

Moisture Regime

Acer negundo L.

mM

Manitoba maple

5 cm

Nutrient Regime

A B C D E

Moisture Regime

TREE

Betula papyrifera Marsh.
white birch

Nutrient Regime

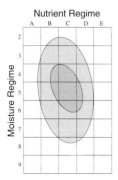

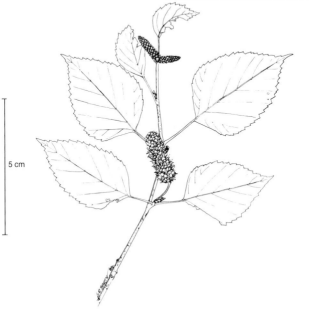

5 cm

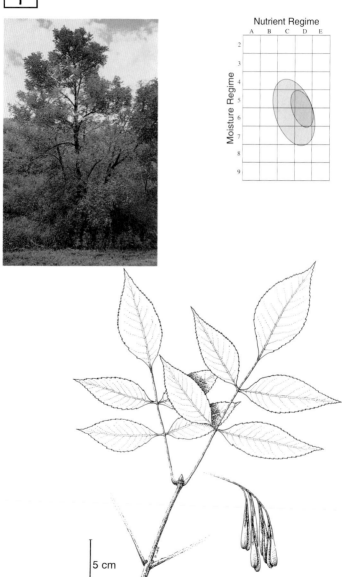

Nutrient Regime

Moisture Regime

5 cm

Larix laricina (Du Roi) K. Koch
tamarack

5 cm

Nutrient Regime

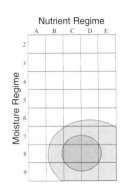

Moisture Regime

5 cm

Nutrient Regime

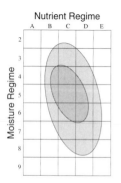

Moisture Regime

Picea mariana (Mill.) B.S.P.
black spruce

5 cm

Nutrient Regime

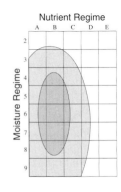

Moisture Regime

5 cm

Nutrient Regime

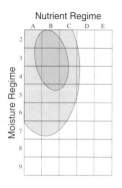

Moisture Regime

Populus balsamifera L.
balsam poplar

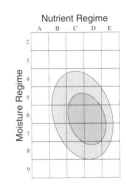

5 cm

Nutrient Regime

A B C D E

Moisture Regime

Populus tremuloides Michx.   tA

aspen

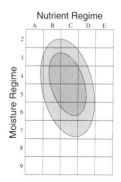

Nutrient Regime

Moisture Regime

5 cm

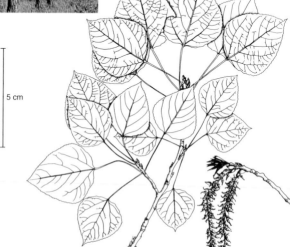

Ulmus americana L.
white elm

Nutrient Regime

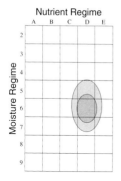

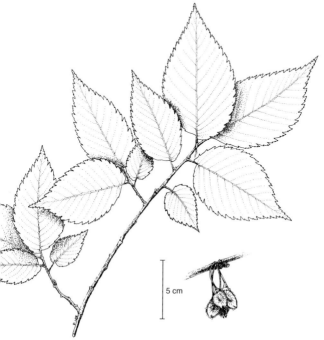

5 cm

Acer spicatum Lamb.
mountain maple

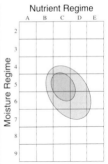

Nutrient Regime

Moisture Regime

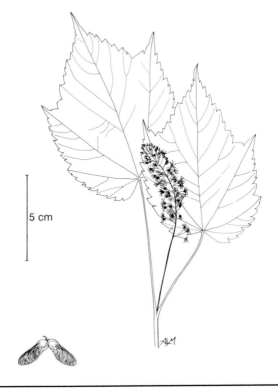

5 cm

Alnus crispa (Ait.) Pursh
green alder

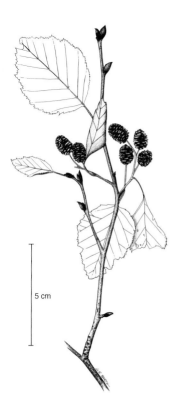

5 cm

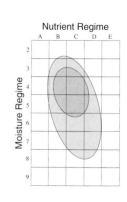

Nutrient Regime

Moisture Regime

Alnus tenuifolia Nutt.
river alder

Nutrient Regime

Moisture Regime

SHRUB

5 cm

saskatoon

Nutrient Regime

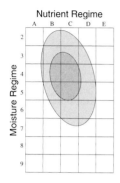

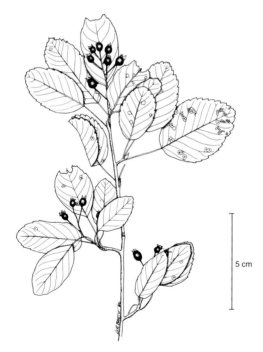

5 cm

Andromeda polifolia L.
bog rosemary

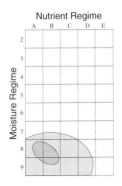

Nutrient Regime

Moisture Regime

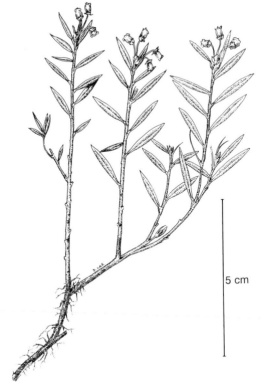

5 cm

SHRUB

Arctostaphylos uva-ursi (L.) Spreng.
bearberry

5 cm

Nutrient Regime

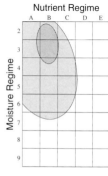

Moisture Regime

Betula pumila L., *Betula glandulosa* Michx.
dwarf birch

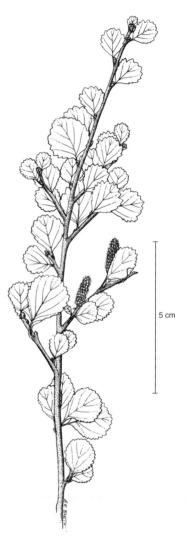

5 cm

Nutrient Regime

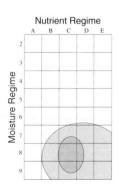

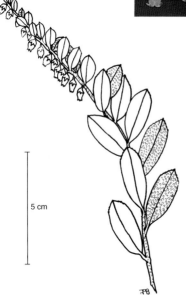

5 cm

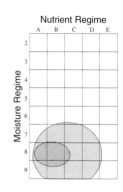

Nutrient Regime

Moisture Regime

Cornus stolonifera Michx.
dogwood

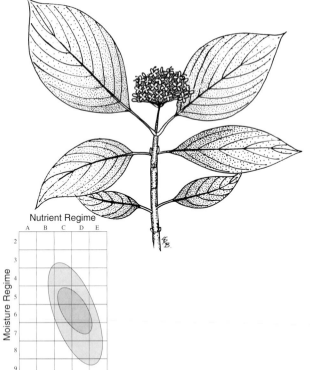

5 cm

Nutrient Regime

Moisture Regime

Corylus cornuta Marsh.
beaked hazelnut

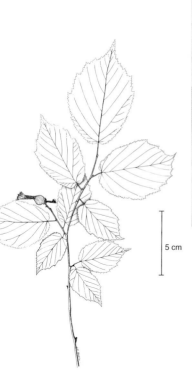

5 cm

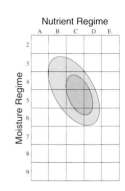

Nutrient Regime

Moisture Regime

SHRUB

11-23

Diervilla lonicera Mill.
bush honeysuckle

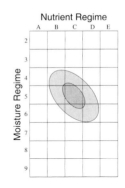

Nutrient Regime

Moisture Regime

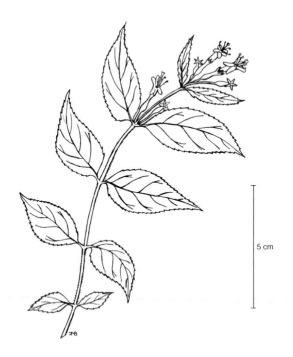

5 cm

Hudsonia tomentosa Nutt.
sand heather

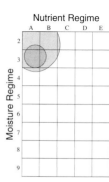

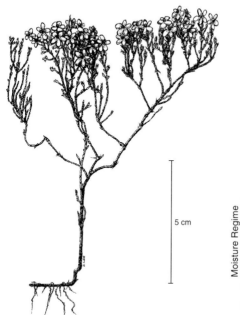

5 cm

Nutrient Regime

Moisture Regime

Kalmia polifolia Wang.
northern laurel

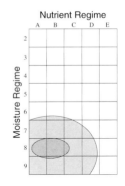

Nutrient Regime

Moisture Regime

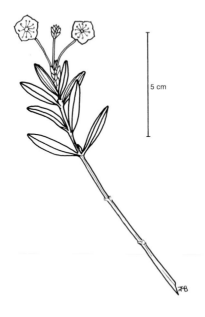

5 cm

SHRUB

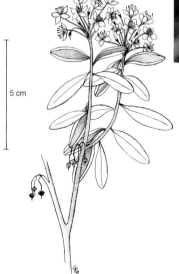

5 cm

Nutrient Regime

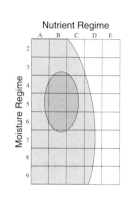

Linnaea borealis L.
twin-flower

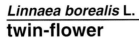

Nutrient Regime

Moisture Regime

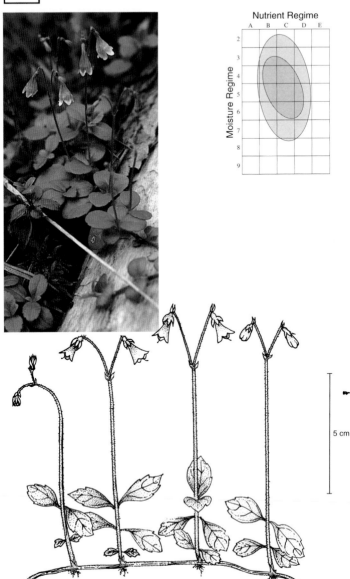

5 cm

SHRUB

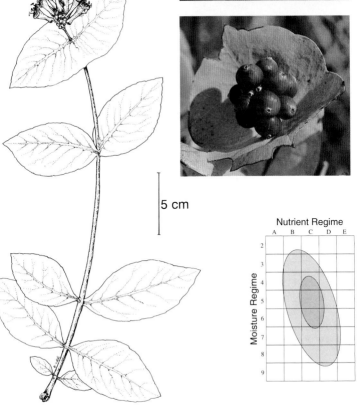

5 cm

Nutrient Regime

Moisture Regime

Lonicera involucrata (Richards.) Banks
bracted honeysuckle

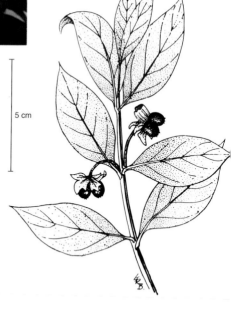

5 cm

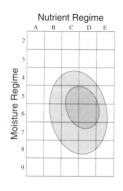

Nutrient Regime

Moisture Regime

Oxycoccus microcarpus Turcz.
small bog cranberry

Nutrient Regime

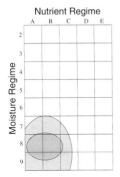

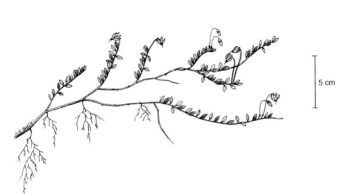

5 cm

SHRUB

11-31

Prunus pensylvanica L.f.
pin cherry

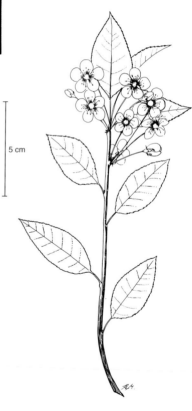

5 cm

Nutrient Regime

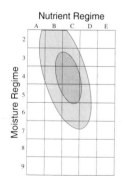

SHRUB

Nutrient Regime

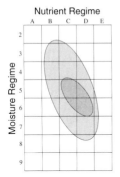

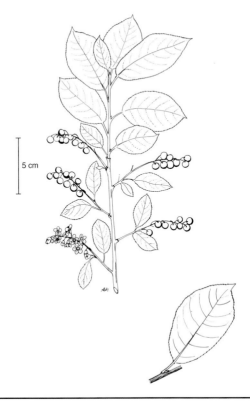

5 cm

Rhamnus alnifolia L'Hér.
alder-leaved buckthorn

Nutrient Regime

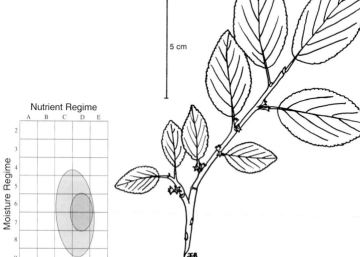

5 cm

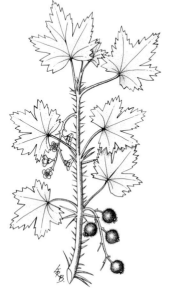

5 cm

Nutrient Regime

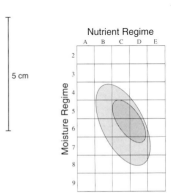

Ribes triste Pall.
wild red currant

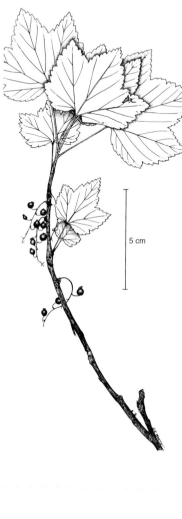

5 cm

Nutrient Regime

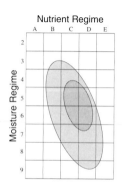

prickly rose

Nutrient Regime

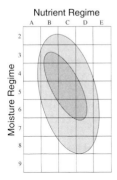

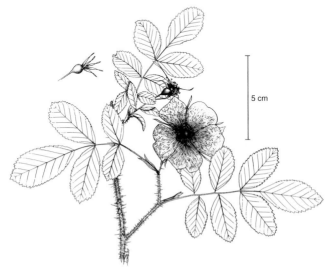

5 cm

Rubus idaeus L.
wild red raspberry

1 cm

5 cm

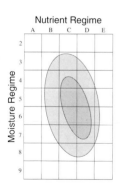

Nutrient Regime

A B C D E

Moisture Regime

SHRUB

Nutrient Regime

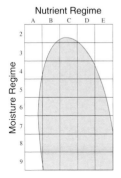

Moisture Regime

5 cm

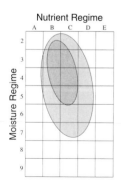

Shepherdia canadensis (L.) Nutt.
Canada buffalo-berry

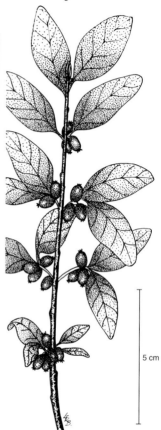

5 cm

Nutrient Regime

A B C D E

Moisture Regime

2
3
4
5
6
7
8
9

SHRUB

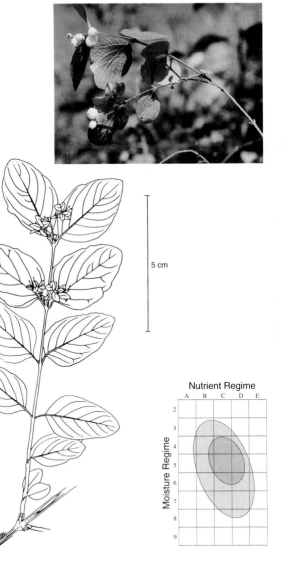

Nutrient Regime

Moisture Regime

5 cm

Vaccinium myrtilloides Michx.
blueberry

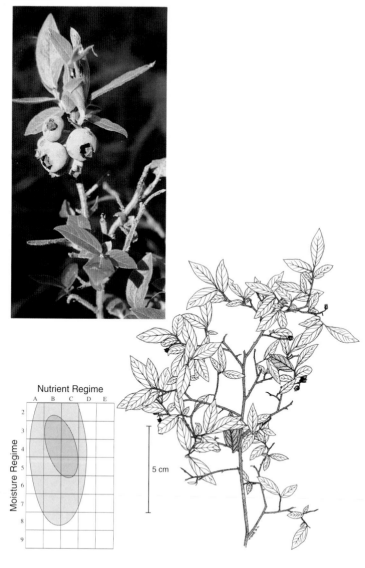

Nutrient Regime

Moisture Regime

5 cm

SHRUB

Vaccinium vitis-idaea L.
bog cranberry

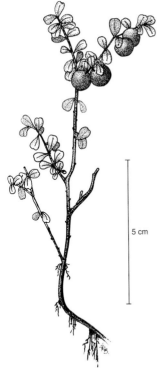

5 cm

Nutrient Regime

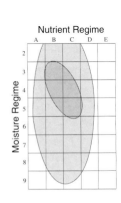

Moisture Regime

low-bush cranberry

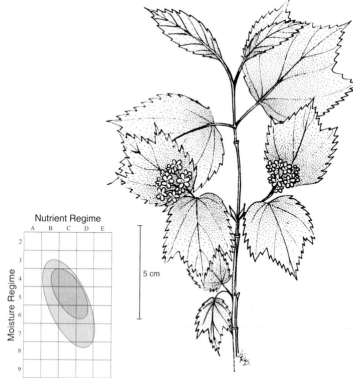

Nutrient Regime

Moisture Regime

5 cm

Viburnum opulus L.
high-bush cranberry

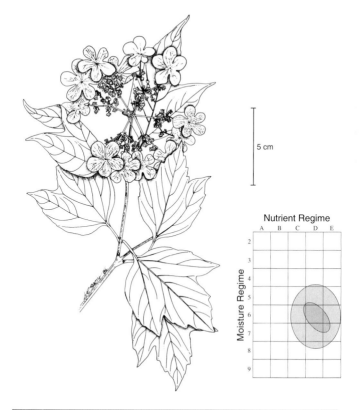

5 cm

Nutrient Regime

Moisture Regime

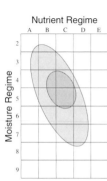

Nutrient Regime

Moisture Regime

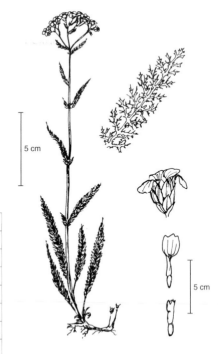

5 cm

5 cm

Nutrient Regime

Moisture Regime

5 cm

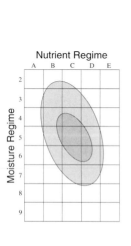

Aster ciliolatus Lindl.
Lindley's aster

Nutrient Regime

5 cm

FORB

Aster conspicuus Lindl.
showy aster

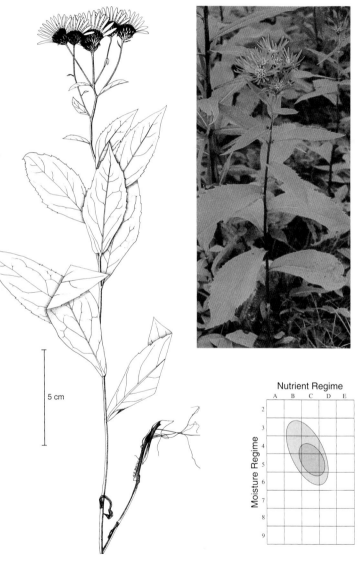

5 cm

Nutrient Regime

Moisture Regime

Athyrium filix-femina (L.) Roth
lady fern

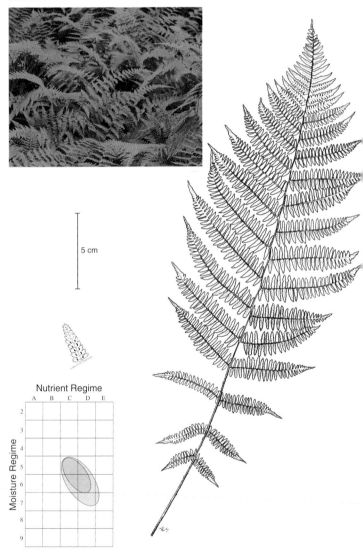

5 cm

Nutrient Regime

Moisture Regime

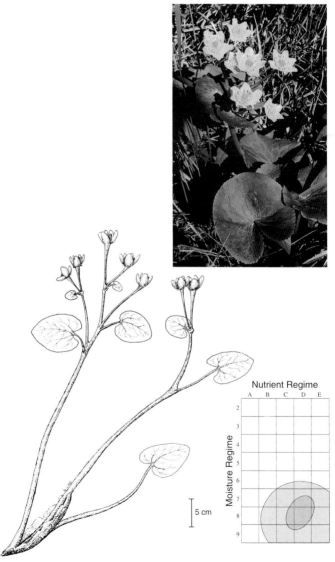

Nutrient Regime

Moisture Regime

Cornus canadensis L.
bunchberry

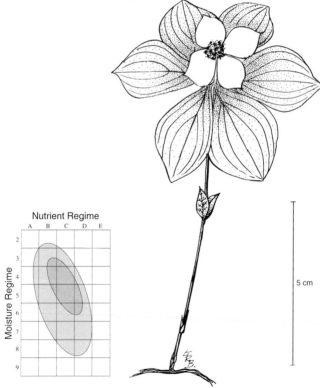

Nutrient Regime

5 cm

FORB

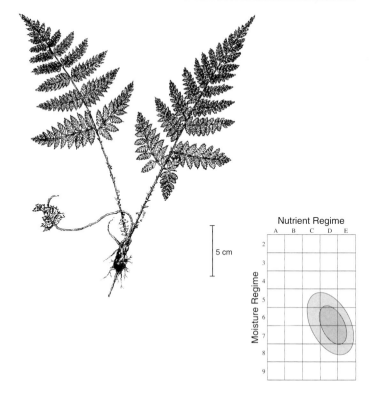

Nutrient Regime

Moisture Regime

5 cm

Epilobium angustifolium L.
fireweed

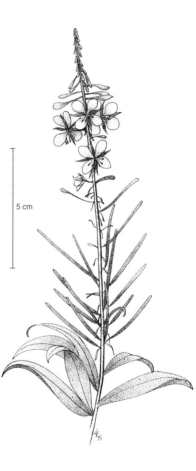

5 cm

Nutrient Regime

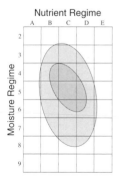

FORB

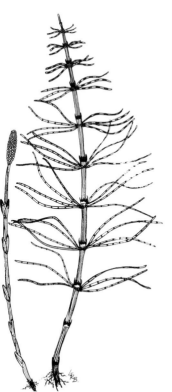

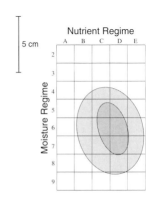

5 cm

Equisetum pratense Ehrh.
meadow horsetail

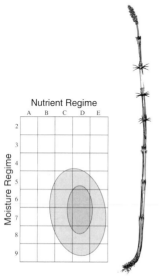

Nutrient Regime

A   B   C   D   E

Moisture Regime

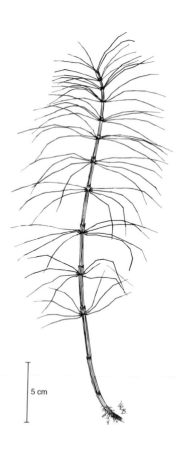

5 cm

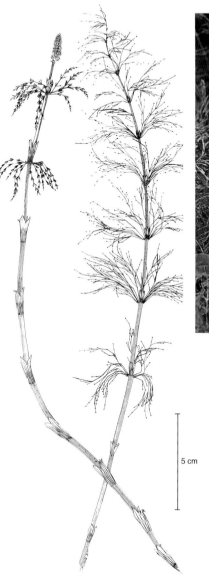

5 cm

Nutrient Regime

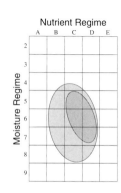

Fragaria vesca L.
woodland strawberry

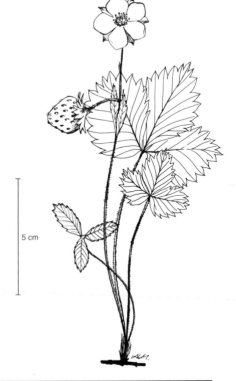

Nutrient Regime

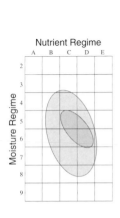

5 cm

FORB

Nutrient Regime

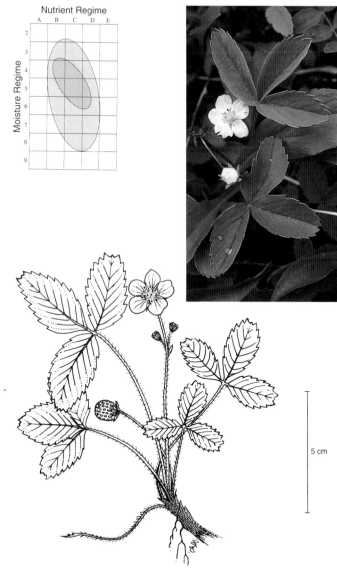

5 cm

Galium boreale L.
northern bedstraw

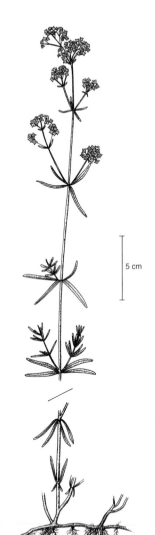

5 cm

Nutrient Regime

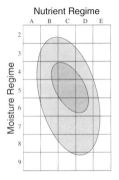

Moisture Regime

Galium triflorum Michx.
sweet-scented bedstraw

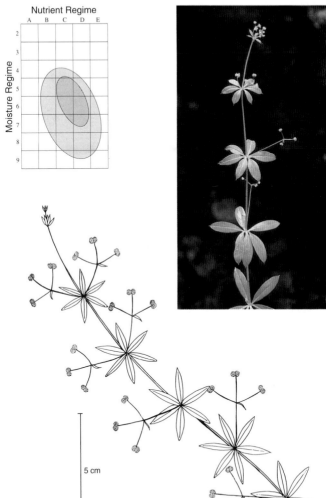

Nutrient Regime

Moisture Regime

5 cm

Gymnocarpium dryopteris (L.) Newm.
oak fern

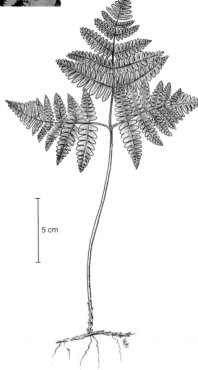

5 cm

Nutrient Regime

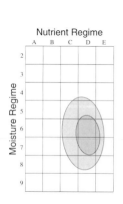

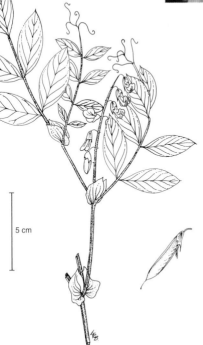

5 cm

Nutrient Regime

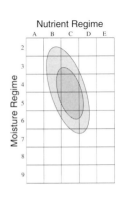

Moisture Regime

Lycopodium annotinum L.
stiff club-moss

5 cm

Nutrient Regime

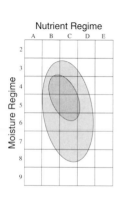

Maianthemum canadense Desf.
wild lily-of-the-valley

5 cm

Nutrient Regime

Moisture Regime

Matteuccia struthiopteris (L.) Todaro
ostrich fern

Nutrient Regime

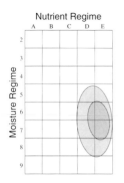

5 cm

FORB

5 cm

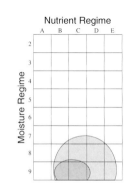

Nutrient Regime

A   B   C   D   E

Moisture Regime

Mertensia paniculata (Ait.) G. Don.
tall lungwort

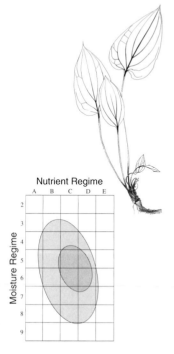

Nutrient Regime

A B C D E

Moisture Regime

5 cm

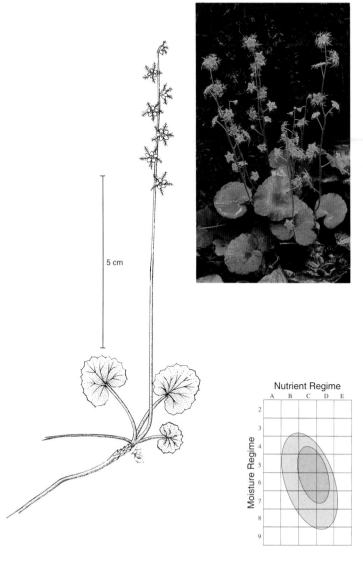

5 cm

Nutrient Regime

Moisture Regime

Orthilia secunda (L.) House
one-sided wintergreen

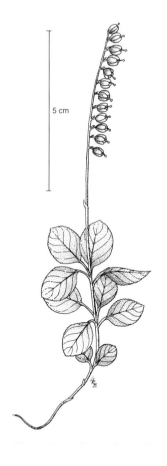

5 cm

Nutrient Regime

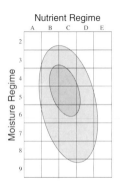

FORB

5 cm

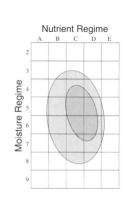

Nutrient Regime

Moisture Regime

Pyrola asarifolia Michx.
common pink wintergreen

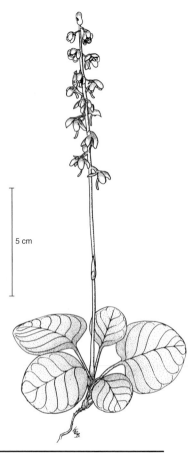

5 cm

Nutrient Regime

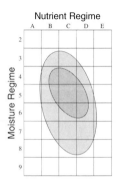

A   B   C   D   E

Moisture Regime

2
3
4
5
6
7
8
9

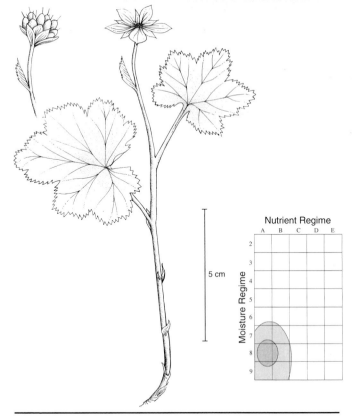

Nutrient Regime

Moisture Regime

5 cm

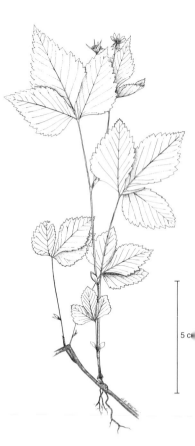

5 cm

Nutrient Regime

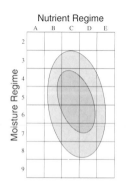

Moisture Regime

5 cm

Nutrient Regime

Moisture Regime

Smilacina trifolia (L.) Desf.
three-leaved Solomon's seal

Nutrient Regime

Moisture Regime

5 cm

5 cm

Nutrient Regime

Moisture Regime

Thalictrum venulosum Trel.
veiny meadow rue

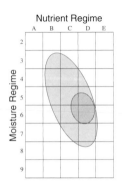

5 cm

Nutrient Regime

A B C D E

Moisture Regime

FORB

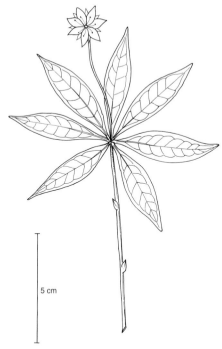

Nutrient Regime

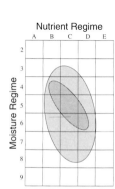

Typha latifolia L.
cattail

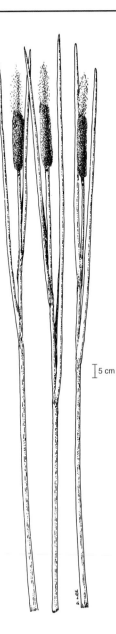

5 cm

Nutrient Regime

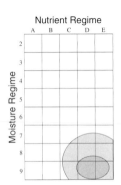

Nutrient Regime

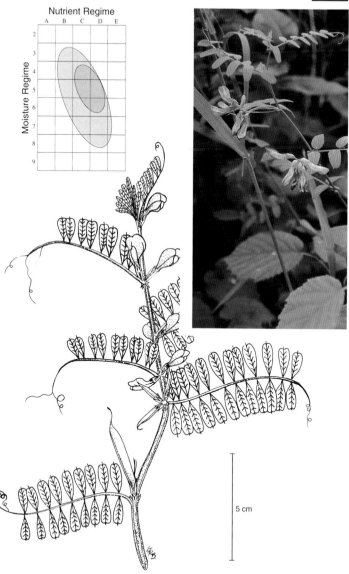

5 cm

Viola canadensis L.
western Canada violet

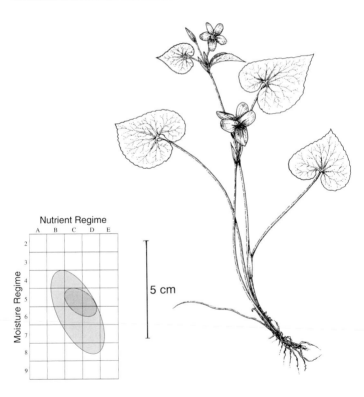

Nutrient Regime

5 cm

FORB

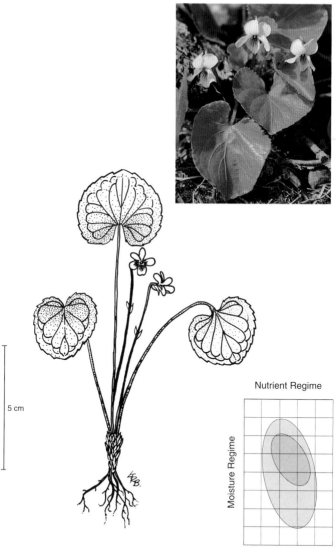

Nutrient Regime

Moisture Regime

5 cm

Nutrient Regime

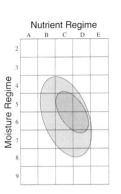

5 cm

Calamagrostis canadensis (Michx.) Beauv.
marsh reed grass

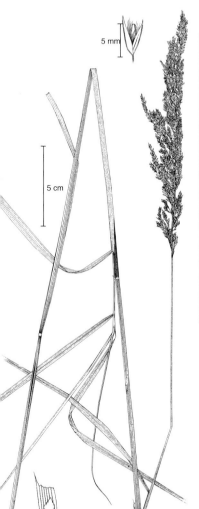

5 mm

5 cm

1 cm

Nutrient Regime

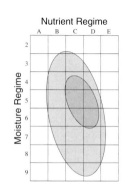

A B C D E

Moisture Regime

2
3
4
5
6
7
8
9

Carex concinna R.Br.
beautiful sedge

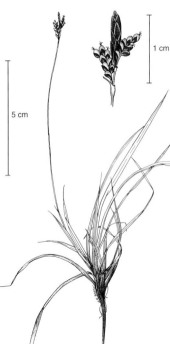

5 cm

1 cm

Nutrient Regime

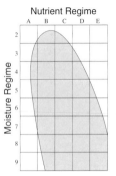

Moisture Regime

Elymus innovatus Beal
hairy wild rye

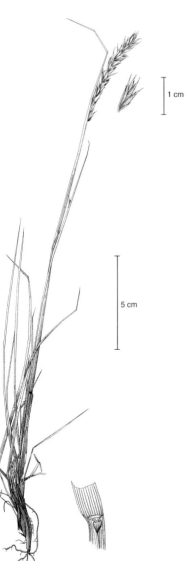

1 cm

5 cm

Nutrient Regime

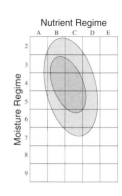

Moisture Regime

	A	B	C	D	E
2					
3					
4					
5					
6					
7					
8					
9					

Juncus balticus Willd.
wire rush

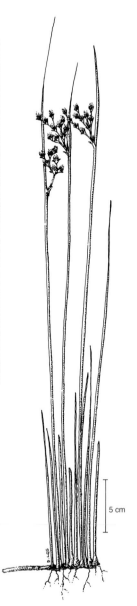

5 cm

Nutrient Regime

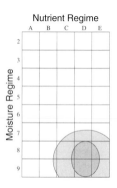

Moisture Regime

GRASS

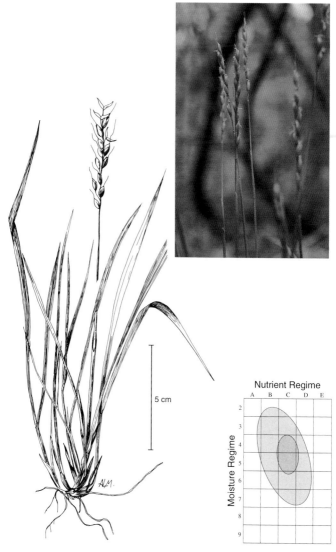

5 cm

Nutrient Regime

A B C D E

Moisture Regime

Phalaris arundinacea L.
reed canary grass

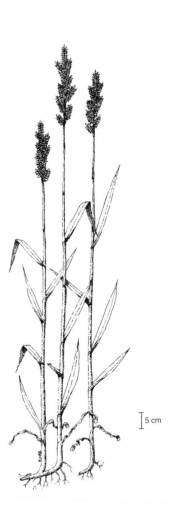

⊺ 5 cm

Nutrient Regime

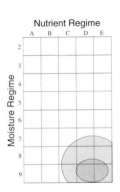

Moisture Regime

Phragmites australis (Cav.) Trin. ex Steud.
giant reed grass

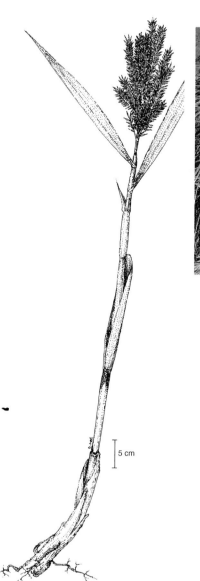

5 cm

Nutrient Regime

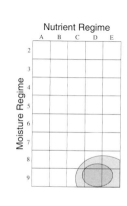

Moisture Regime

A B C D E

Schizachne purpurascens (Torr.) Swallen
false melic

Nutrient Regime

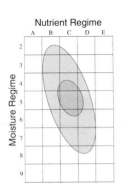

5 cm

I 5 cm

Nutrient Regime

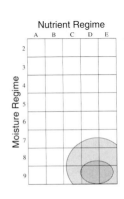

Aulacomnium palustre (Hedw.) Schwaegr.
tufted moss

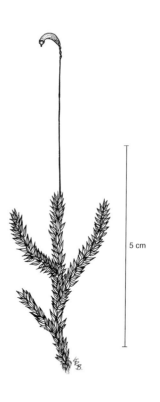

5 cm

Nutrient Regime

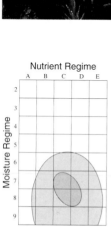

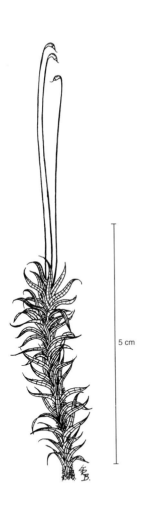

5 cm

Nutrient Regime

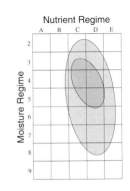

Moisture Regime

Drepanocladus uncinatus (Hedw.) Warnst.
sickle moss (brown moss)

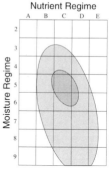

Nutrient Regime

Moisture Regime

5 cm

PDB

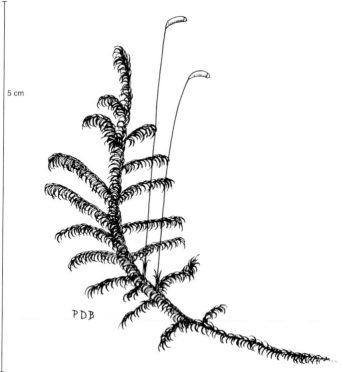

Hylocomium splendens (Hedw.) B.S.G.
stair-step moss

Nutrient Regime

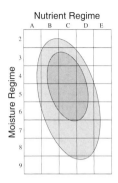

5 cm

Pleurozium schreberi (Brid.) Mitt.
Schreber's moss

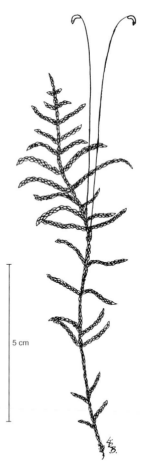

Nutrient Regime

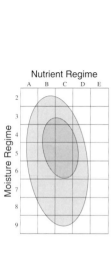

5 cm

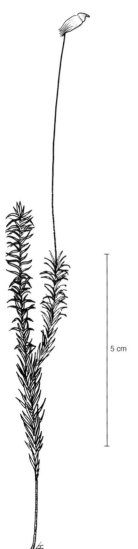

5 cm

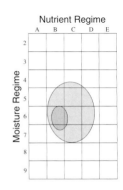

Nutrient Regime

A B C D E

Moisture Regime

Ptilium crista-castrensis (Hedw.) De Not.
knight's plume moss

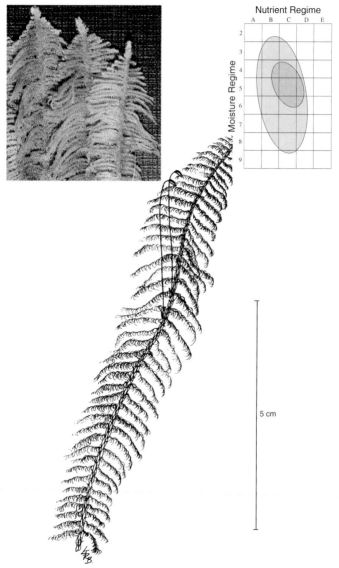

Nutrient Regime

Moisture Regime

5 cm

5 cm

Nutrient Regime

Moisture Regime

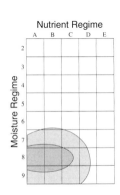

Tomenthypnum nitens (Hedw.) Loeske
golden moss

5 cm

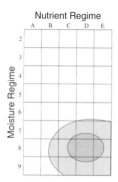

Nutrient Regime

A B C D E

Moisture Regime

Cladina mitis (Sandst.) Hale & W. Culb.
reindeer lichen

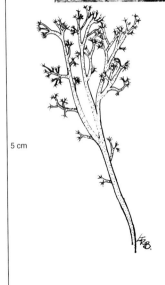

5 cm

Nutrient Regime

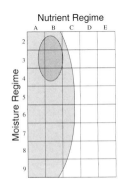

Moisture Regime

Cladonia borealis S. Stenroos
red pixie-cup

5 cm

Nutrient Regime

Moisture Regime

LICHEN

Peltigera aphthosa (L.) Willd.
studded leather lichen

Nutrient Regime

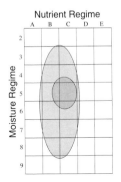

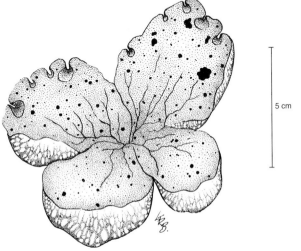

5 cm

12.0 MANAGEMENT INTERPRETATIONS

Management interpretations are interpretations about the limitations and opportunities of the ecological units in the classification system. These have been developed through data analysis, literature review, public workshops, and expert opinion. They present the user with a general outline of limitations and opportunities that together with the user's knowledge and experience should be applied in a creative manner. Some management interpretations will change dramatically with time, season of year, economic conditions, existing technology, scale of application, and program objectives (Still and Utzig 1982). Under no circumstances should the information in this guide be construed as a formal recommendation or guideline for resource management, or as a prescription for specific sites.

Five levels were used for rating the ecosites and soil types—low (L), low to medium (L-M), medium (M), medium to high (M-H), and high (H). The season-of-harvest management interpretation used a rating nomenclature different from what is described above (Section 12.11). All ratings are presented at the bottom of the ecosite and soil type fact sheets.

The relative meaning of an interpretation rating and the variables that were used in the rating process are described in Sections 12.1 to 12.11. All management interpretations are based on the variability of important site and soil characteristics associated with each ecosite and soil type.

Tables 6 and 7 summarize the management interpretation ratings for the soil and ecosite types, respectively.

12.1 Drought limitations

Droughty conditions are associated with rapidly drained soils that draw water away from the rooting zone for a significant portion of the growing season. Typically, sites that are limited by drought are associated with coarse-textured soils or are situated on steep south-facing slopes where insolation levels and evaporation rates are high. Remedial silvicultural efforts such as planting during moist periods of the year as in the spring, using drought-tolerant species (e.g., jack pine), using stock with small

Table 6. Summary of management interpretations for each soil type

	SV1	SV2	SV3	SV4	SD1	SD2	SD3	SD4
Drought limitations	H	H	H	H	M	M	M	M
Excess moisture	L	L	L	L	L	L	L	L
Rutting hazard	L	L	L	L	L	L	L-M	M
Compaction hazard	L	L	L-M	L-M	L	L	M	M
Puddling hazard	L	L	M	M	L	L	M	H
Erosion hazard	L	L-M	H	H	L	L-M	H	H
Frost heave hazard	L	L	L-M	L-M	L	L-M	M	M
Soil temperature limitations	L	L	L	L	L	L	L	L
Vegetation competition	L	L	L	L	M	M	M	M
Windthrow hazard	L-M	L	L	L	L-M	L	L	L
Productivity	L	L	L	L	M	M	M	M
Season of harvest	W-Sp	W-Sp	A	A	A	A	A	A

Table 6.　concluded

	SM1	SM2	SM3	SM4	SMp	SWm	SWp	SR	SS
Drought limitations	L	L	L	L	L	L	L	L	-[a]
Excess moisture	L-M	L-M	L-M	L-M	M	H	H	H	-
Rutting hazard	L	L-M	M-H	M-H	H	H	H	H	-
Compaction hazard	L	L-M	M-H	H	H	H	H	L	-
Puddling hazard	L	L	M-H	H	H	H	H	L	-
Erosion hazard	L	M	M-H	M-H	H	H	H	L	-
Frost heave hazard	L	L-M	M-H	M-H	H	H	H	H	-
Soil temperature limitations	L-M	L-M	L-M	L-M	M-H	H	H	H	-
Vegetation competition	M-H	M-H	H	H	H	M-H	M-H	L	-
Windthrow hazard	L-M	L	L	L	M-H	H	H	H	-
Productivity	H	H	H	H	H	L-H	L	L	-
Season of Harvest	A	A	A	A	W	W	W	W	-

[a]SS soils were not sampled but are expected to occur on the northern edge of the study area bordering the Boreal Shield Ecozone.

Table 7. Summary of management interpretations for each ecosite

	a	b	c	d	e	f	g	h	i	j	k	l	m
Drought limitations	H	M-H	L-H	L-M	L	L	L	L	L	L	L	L	L
Excess moisture	L	L	L	L	M-H	M-H	M-H	H	H	H	H	H	H
Rutting hazard	L	L	L-M	L	H	H	H	H	H	H	H	H	H
Compaction hazard	L	L-M	M	M	H	H	H	H	M	L	L	L	L
Puddling hazard	L	L	L-M	M	H	H	H	H	M	L	L	L	L
Erosion hazard	L	L	M	M-H	H	H	H	H	H	L	L	L	H
Frost heave hazard	L	L-M	L-M	M	H	H	H	H	H	H	H	H	n/a[a]
Soil temperature limitations	L	L	L	L	M	M	H	H	H	H	H	H	n/a
Vegetation competition	L	M	L	H	H	H	M	H	H	L	L	L	n/a
Windthrow hazard	L	L-M	L	L	L-M	L-M	H	H	n/a	H	H	H	n/a
Productivity	L	M	L-M	M-H	H	H	L-M	H	n/a	L	L	L	n/a
Season of harvest	W-Sp	A	A	A	W	W	W	W	n/a	n/a	n/a	n/a	n/a

[a] n/a = not applicable.

tops and large root systems, and utilizing micro-shelter planting sites can all help alleviate the effects of drought (Strong and Carnell 1995).

Ratings are based on the moisture regime of the ecosites and soil types. A high drought limitation rating indicates severe limitations, while low ratings indicate few or no limitations.

12.2 Excess moisture

Excess soil moisture is a concern because serious site degradation can occur if proper management practices are not followed. Operating heavy equipment on wet sites can cause serious rutting, compaction, and puddling damage (Braumandl and Curran 1992) and therefore should be avoided. Winter months are a suitable period for operating on wet sites as the ground is frozen and snow cover acts as a disturbance buffer (Corns and Annas 1986; Braumandl and Curran 1992). Other management practices exist to address the concerns associated with wet sites. All options should be explored.

Ratings are based on the moisture regime of the ecosites and soil types. A high excess moisture rating indicates severe limitations, while low ratings indicate little or no limitations.

12.3 Rutting and compaction hazard

Traffic on soil by machines most often modifies soil quality through compaction, remoulding, puddling, and/or soil displacement which in turn affects several interrelated soil physical properties. The modification that predominates depends on soil wetness, applied stress, and number of passes (McNabb 1995). Soil texture may also be important, especially when soils are at moisture levels close to field capacity (Hausenbuiller 1985; McKee et al. 1985).

The risk of causing soil compaction or rutting by forestry operations should be evaluated before beginning operations as both risks are greatly influenced by the amount of water in the soil at the time of disturbance. Risk assessments are based on soil water content and on estimates of the time it takes a wet soil to drain (Alberta Forest Products Association/ Land and Forest Service 1995).

The rating system presented here does not replace the operational assessment but is designed as a planning tool. It can be used as part of the decision process when evaluating whether an area has the potential for supporting operations in the summer months.

Soil modifications affect four physical soil processes important to an organism's health: water supply and flux, heat flux, soil strength, and gas diffusion (McNabb 1995). Simply stated, the effects of compaction and rutting are manifested in changes in water infiltration rates, soil heat flux, root penetration, and oxygen supply in the soil. All of these conditions may influence soil quality and ultimately soil productivity.

The rating system is based primarily on moisture regime and related soil drainage with soil texture considered for coarse-textured soils (less than 20% silt and clay). High risk ratings indicate that it is unlikely that summer operations would be possible, medium ratings indicate that operations may be possible in dry periods while sites with low risk ratings are good candidates for summer and fall operations. Current moisture conditions should always be evaluated before initiating operations.

12.4 Soil puddling hazard

Soil puddling is the process by which the structure of the surface soil layer is destroyed through the realignment of clay particles, ultimately leading to restricted drainage (Pritchett 1979; Hausenbuiller 1985; McKee et al. 1985). The operation of heavy equipment and the impact of rain on exposed mineral surfaces can cause surfaces to form crusts. (Hausenbuiller 1985; Corns and Annas 1986). Wet, fine-textured soils are most susceptible to puddling (Corns and Annas 1986).

Seedling mortality rates on puddled soils tend to be higher than on undisturbed soils because the dense crust that forms tends to reduce aeration properties of the soil (Hausenbuiller 1985; Corns and Annas 1986). The best prevention against soil puddling is to avoid operations during wet periods and to leave the surface organic layer intact (Corns and Annas 1986). Low impact machinery should be considered where a moderate risk of soil puddling occurs (Racey et al. 1989).

Ratings were based on the moisture regime and surface texture of the ecosites and soil types and on the assumption that organic layers are disturbed during operations. Reduce the soil puddling hazard rating by one level (e.g., high to medium) if the surface organic layers are not disturbed.

12.5   Soil erosion hazard

Ecosites and soil types were rated for surface erosion hazard caused by water. Infiltration capacity and structural stability are central to evaluating erosion hazard because they are regarded as the most important factors in controlling water erosion (Buckman and Brady 1960). Numerous soil and site variables affect infiltration capacity and structural stability including the extent and type of vegetation cover, the thickness of the LFH layer, the type of humus form, texture of the surface and C horizons, degree of carbonate cementing, coarse fragment content, slope angle, and length of slope (Comeau et al. 1982; Corns and Annas 1986; Racey et al. 1989). Climatic factors such as rainfall intensity, duration and seasonal distribution, and the rapidity of snow melt affect erosion, but are difficult to relate to a particular ecosite or soil type. Soil erosion hazard decreases as clay or sand content increases, and increases as percent silt increases (Wischmeier and Meyers 1973). As organic matter depth and vegetation cover increase, erosion hazard decreases.

Ratings were based on the moisture regime and surface texture of the ecosites and soil types and on the assumption that organic layers are disturbed during operations. Reduce the soil erosion hazard rating by one level (e.g., high to medium) if the surface organic layers and/or understory vegetation layers are not disturbed.

12.6   Frost heave hazard

When soil water segregates and freezes into a layer or a lens of ice near the soil surface, frost heaving of the seedlings occurs. Frost heave hazard is the risk of newly planted trees being forced out or partly out of the soil as ice lens formation causes an uplift of the surface soil.

The most susceptible soils are those with intermediate permeability and water tension (Heidmann 1976). Sandy soils are very permeable but exhibit low tension due to larger air spaces between particles. Conversely, clay soils exhibit high tension but low permeability (Racey et al. 1989). Fine-textured surface horizons with a high silt content and imperfect drainage are especially susceptible. Topographic position can be an important determinant of frost heave susceptibility, with depressions being most susceptible (Corns and Annas 1986). Frost heave hazard may be reduced by maintaining brush and ground cover, or an intact organic layer around tree seedlings (Singh 1976; Comeau et al. 1982; Corns and Annas 1986).

Ratings were based on moisture regime and surface texture, and on the assumption that organic layers are disturbed during operations. Reduce the frost heave hazard rating by one level (e.g., high to medium) if the surface organic layers are not disturbed.

12.7   Soil temperature limitation

Soil temperature is important as it relates to seedling growth and survival. The rate of root development and the ability of plants to uptake water is considerably less in cold soils than in warm soils (Larcher 1983). Thus, seedlings planted in cold soils are disadvantaged during the critical establishment period. Areas where cold soils are prevalent include depressions, north-facing slopes (300° to 60° aspect) greater than 30%, sites located at the base of major slopes, and in valleys (Strong and Carnell 1995). Opportunities exist to increase soil temperatures to more favorable levels using various site preparation methods that loosen and expose mineral soil to the sun. Educating tree planters to plant in idealized micro-site locations will also help increase the survival rates of seedlings.

Ratings were based on moisture regime, topographic position, and surface texture and on the assumption that organic layers are disturbed during operations. Increase the rating by one level (e.g., medium to high) if organic layers are not disturbed. On southerly aspects, where temperatures are otherwise too high, organic layer presence may be beneficial.

12.8   Vegetation competition

Assessing the degree of vegetation competition associated with each ecosite and soil type is important as it relates to forestry operations and planning such as choosing an appropriate planting stock, site preparation methods, and projected management costs. Research and experience has shown that vegetation competition is associated with the height and percent cover of shrubs, forbs, and grasses and whether a seedling is overtopped by a competitor (Geographic Dynamics Corp. 1995a). Some of the more competitive species include shrubs such as green alder (*Alnus crispa*), river alder (*Alnus tenuifolia*), willow (*Salix* spp.), and bracted honeysuckle (*Lonicera involucrata*), tall or prolific forbs such as fireweed (*Epilobium angustifolium*) and wild sarsaparilla (*Aralia nudicaulis*), and grasses such as hairy wild rye (*Elymus innovatus*) and most particularly marsh reed grass (*Calamagrostis canadensis*) (adapted from Corns and Annas 1986). Marsh reed grass is considered to be one of the more highly competitive species with conifer seedlings (Corns and Annas 1986).

Ratings were based on moisture regime, nutrient regime, and surface texture associated with the ecosites and soil types and on the assumption that organic layers are disturbed during operations. In general, high ratings were assigned to those ecosites and soil types that are moist and nutrient rich. Low ratings were assigned to ecosites and soil types that are very dry, rapidly drained, and nutrient poor where dense understory vegetation is uncommon.

12.9   Windthrow hazard

Several environmental and man-made factors influence the susceptibility of a site to windthrow hazard. These factors include exposure, cut block layout, and topography (Braumandl and Curran 1992) and should always be taken into consideration when assessing the windthrow hazard of a particular site. Shallow root systems evident on sites with thick organic layers or high water tables increase the chance of windthrow (Braumandl and Curran 1992) while coarse-textured soils can reduce the ability of a root system to anchor trees firmly (Strong and Cornell 1995).

Windthrow hazard ratings were based on organic matter thickness, presence of water table, tree-rooting habit, and effective soil texture.

12.10 Productivity

Tree productivity ratings were assessed for all ecosites with forested phases. Ratings for the ecosites are relative and are based on the site index at 50 years BH age of two primary species and on mean annual increment (MAI) (Table 8). Mean annual increment estimates for each

Table 8.   Productivity rating of the forested ecosites

Ecosite	Site index species 1	Site index species 2	Mean annual increment (m³/ha/yr)	Productivity rating
a lichen	jP 14.9	–[a] –	1.6	L
b blueberry	tA 18.3	jP 18.2	2.8	M
c Labrador tea–submesic	jP 16.1	bS 12.2	2.7	L-M
d low-bush cranberry	tA 20.0	wS 19.7	3.8	M-H
e dogwood	tA 21.3	wS 18.5	3.4	H
f ostrich fern	bP 22.3	wS 23.6	3.1	H
g Labrador tea–hygric	bS 12.9	jP 15.5	2.0	L-M
h horsetail	wS 18.1	bP 15.0	3.3	H
j bog	bS 8.0	– –	0.8	L
k poor fen	bS 10.7	tL 13.2	1.0	L
l rich fen	tL 13.5	– –	1.1	L

[a] not applicable.

ecosite were based on a weighted average of the total MAI estimates for each phase within that ecosite.

It is difficult to rate the productivity of ecosites strictly on site index and mean annual increment since the primary tree species of each ecosite may vary and MAI estimates have not been adjusted to biological rotation age. To adjust for the error inherent in these calculations, our understanding of the effects of soil moisture, soil texture, and nutrient availability on productivity were also criteria used to rate the ecosites and soil types.

For the boreal mixedwood forest in Saskatchewan, Kabzems (1971) found that differences in the MAI of white spruce stands at rotation age were correlated with variations in soil texture and drainage. He found that clay loam, moderately well-drained soils were the most productive. These general trends were not strong when using the soil type classification system developed for this field guide due in part to the small sample size. Sample size or confidence in prediction was increased by calculating site index by soil moisture class (very dry, dry, moist, and wet). High site index values were typically associated with moist soils, moderate values with dry soils, and the lowest values with very dry soils. Site index associated with wet soils varied considerably depending on tree species. When soils were grouped by texture class, organic soils (previously grouped with wet soils by moisture class) were associated with the lowest site index values. Based on these trends, interpretations of the productivity of the soil types were developed.

Ratings with high values are indicative of sites with minimal limitations to productivity. Low ratings indicate that there are one or more severe limiting factors.

12.11 Season of harvest

A critical factor to consider when assessing the season of harvest is the amount of moisture associated with a particular ecosite or soil type. Operating machinery on sites with excess moisture will seriously degrade the soil (Corns and Annas 1986). Most sites can be harvested in the winter (W) if soils are frozen. Site and soil conditions should be carefully investigated before harvesting in summer months. Those sites that can be harvested during the summer months can generally be harvested in the winter (A = winter and summer months) (Corns and Annas 1986). Spring harvesting (Sp) is not recommended on any site unless moisture conditions are at acceptable levels. Winter harvesting should be the only option for wet sites as site degradation potential is high. Regardless of the typical moisture conditions associated with an ecosite or soil type, areas should be managed on a site-by-site basis as seasonal variations in precipitation can dramatically affect moisture conditions.

Besides annual variations in precipitation, wildlife considerations may also dictate the timeframe of harvest beyond what is considered normal for a particular ecosite. For example, if caribou are known to browse on a lichen ecosite (a) in a particular harvest area, operations may be limited to winter months so that lichen cover is minimally disturbed[1].

Season-of-harvest ratings are based primarily on soil moisture conditions.

12.12 Plant and wildlife management interpretations

12.12.1 Plant species richness and diversity

Ecosystems vary in the number and quantity of each species they contain. There are three common measures of species biodiversity–species richness, species evenness, and species diversity. Species richness is simply the number of species in a given area or community, which is expressed in this guide as the number of species/sample plot. One problem with species richness is that it treats rare and common species as equal. The distribution of individuals among the species is called species evenness, or species equability. Evenness is at a maximum when all species have the same number of individuals or the same percent cover. Species diversity combines the concept of species richness and evenness. It is species richness weighted by relative species abundance.

---

[1] Letter from Genny Greif, Saskatchewan Environment and Resource Management, Prince Albert, Saskatchewan, to the Canadian Forest Service, Prince Albert, Saskatchewan, February 16, 1995.

The number of vascular plant species were tallied for each ecosite phase and divided by the number of sample plots to give richness expressed as the mean number of species per sample plot. This information is presented on the ecosite phase fact sheets beside the heading richness. Table 9 lists species richness data by stratum for each ecosite phase. The standard error of the mean and the number of samples used in each calculation are presented in Table 9 and on the ecosite phase fact sheet.

Species diversity was calculated using the Shannon-Wiener index formula:

$$H' = \sum_{i=1}^{s} (p_i)(\log_2 p_i)$$

where: H' = index of species diversity (bits/individual)

p_i = proportion of total percent cover belonging to each species

Species diversity was calculated for all vascular plant species in each plot of each ecosite phase and is presented as the mean diversity on the ecosite phase fact sheets beside the heading diversity. Table 9 lists species diversity data by stratum for each ecosite phase.

Although the Shannon-Wiener index places more emphasis on rare species than other indices of diversity such as Simpson's index (Barbour et al. 1987) it is important to identify which plant species are rare and where they are located. The Saskatchewan Conservation Data Centre tracks biodiversity at two levels: species and ecological community. Elements of biological diversity are ranked to help priorize conservation efforts. Element status can also be compared between global, national, and subnational (or provincial) levels. Table 10 lists rare and endangered plant species found in the Mid-Boreal ecoregions and their global (GRANK) and subnational (SRANK) element ranks. Comments have been provided in Table 10 to help define the probable habitats of these plants.

12.12.2 Wildlife

Wildlife species are integral components of ecosystems. A particular species or guild has specific requirements for space, food, water, thermal regulation, and protection. Various ecological units of the classification

Table 9. Richness and diversity indices for vegetation strata

	Richness	Diversity	Richness	Diversity
	a1 lichen jP (n = 47)		**b4 blueberry wS(jP)** (n = 10)	
TREE	1.30 ± 0.08	0.11 ± 0.04	2.40 ± 0.27	0.74 ± 0.18
SHRUB	6.77 ± 0.38	1.90 ± 0.09	8.30 ± 0.79	1.90 ± 0.20
FORB	6.28 ± 0.61	2.08 ± 0.13	8.10 ± 1.09	2.50 ± 0.23
GRASS	1.91 ± 0.20	0.80 ± 0.11	1.10 ± 0.38	0.30 ± 0.21
MOSS	3.47 ± 0.34	0.90 ± 0.11	2.80 ± 0.63	0.68 ± 0.19
LICHEN	5.43 ± 0.67	0.92 ± 0.10	1.70 ± 0.52	0.51 ± 0.26
Total	*25.15 ± 1.19*	*2.87 ± 0.10*	*24.40 ± 2.36*	*2.89 ± 0.14*
	b1 blueberry jP-tA (n = 32)		**c1 Labrador tea–submesic jP-bS** (n = 44)	
TREE	2.91 ± 0.19	1.19 ± 0.07	1.82 ± 0.11	0.45 ± 0.07
SHRUB	9.50 ± 0.55	2.33 ± 0.11	6.95 ± 0.34	1.95 ± 0.10
FORB	10.78 ± 0.88	2.63 ± 0.12	7.59 ± 0.62	2.38 ± 0.14
GRASS	1.63 ± 0.17	0.52 ± 0.10	1.00 ± 0.13	0.26 ± 0.07
MOSS	4.66 ± 0.37	1.23 ± 0.13	4.18 ± 0.24	0.62 ± 0.06
LICHEN	1.34 ± 0.29	0.47 ± 0.12	2.34 ± 0.39	0.83 ± 0.15
Total	*30.81 ± 1.42*	*3.47 ± 0.09*	*23.89 ± 1.02*	*2.64 ± 0.07*
	b2 blueberry tA(wB) (n = 25)		**d1 low-bush cranberry jP-bS-tA** (n = 11)	
TREE	2.00 ± 0.23	0.33 ± 0.09	2.64 ± 0.36	0.90 ± 0.19
SHRUB	9.56 ± 0.51	2.28 ± 0.12	10.73 ± 0.76	2.44 ± 0.23
FORB	11.88 ± 0.90	2.70 ± 0.17	15.55 ± 1.22	2.65 ± 0.23
GRASS	2.36 ± 0.26	0.91 ± 0.14	1.64 ± 0.31	0.58 ± 0.20
MOSS	1.92 ± 0.33	0.67 ± 0.16	4.27 ± 0.65	1.25 ± 0.17
LICHEN	0.24 ± 0.10	0.04 ± 0.04	0.73 ± 0.36	0.29 ± 0.19
Total	*27.96 ± 1.26*	*3.06 ± 0.11*	*35.55 ± 1.47*	*3.56 ± 0.15*
	b3 blueberry tA-wS (n = 12)		**d2 low-bush cranberry tA** (n = 123)	
TREE	2.67 ± 0.22	1.08 ± 0.09	1.98 ± 0.09	0.39 ± 0.04
SHRUB	10.50 ± 0.88	2.56 ± 0.12	10.66 ± 0.26	2.37 ± 0.05
FORB	11.08 ± 1.11	2.78 ± 0.17	16.64 ± 0.39	3.14 ± 0.05
GRASS	2.17 ± 0.30	0.84 ± 0.20	2.57 ± 0.13	0.98 ± 0.07
MOSS	2.75 ± 0.46	0.85 ± 0.19	1.80 ± 0.18	0.69 ± 0.07
LICHEN	1.33 ± 0.64	0.33 ± 0.19	0.59 ± 0.12	0.20 ± 0.05
Total	*30.50 ± 1.62*	*3.58 ± 0.10*	*34.24 ± 0.68*	*3.35 ± 0.05*

Table 9.   **continued**

	Richness	Diversity	Richness	Diversity
d3 low-bush cranberry tA-wS (n=136)			**e3 dogwood wS** (n = 15)	
TREE	2.98 ± 0.08	1.23 ± 0.03	2.67 ± 0.30	0.73 ± 0.12
SHRUB	9.27 ± 0.25	2.38 ± 0.05	8.67 ± 0.86	2.14 ± 0.21
FORB	15.52 ± 0.43	3.07 ± 0.05	16.33 ± 1.08	3.19 ± 0.16
GRASS	1.53 ± 0.08	0.54 ± 0.05	2.33 ± 0.48	0.86 ± 0.24
MOSS	4.36 ± 0.29	1.21 ± 0.07	6.73 ± 0.89	1.74 ± 0.17
LICHEN	1.38 ± 0.17	0.54 ± 0.07	2.00 ± 0.72	0.75 ± 0.29
Total	*35.04 ± 0.79*	*3.49 ± 0.04*	*38.73 ± 2.43*	*3.54 ± 0.16*
d4 low-bush cranberry wS (n=31)			**f1 ostrich fern bP-tA** (n = 5)	
TREE	2.19 ± 0.19	0.36 ± 0.06	2.20 ± 0.49	0.68 ± 0.22
SHRUB	6.74 ± 0.48	1.76 ± 0.14	9.80 ± 1.24	1.21 ± 0.36
FORB	13.03 ± 0.93	3.04 ± 0.09	18.20 ± 1.69	2.74 ± 0.45
GRASS	0.87 ± 0.17	0.22 ± 0.09	2.40 ± 0.81	0.74 ± 0.34
MOSS	6.03 ± 0.63	1.31 ± 0.09	3.60 ± 0.81	1.38 ± 0.37
LICHEN	1.87 ± 0.37	0.73 ± 0.18	0.20 ± 0.20	0.00 ± 0.00
Total	*30.74 ± 1.62*	*2.91 ± 0.10*	*36.40 ± 2.96*	*3.07 ± 0.22*
e1 dogwood bP-tA (n = 29)			**f2 ostrich fern mM-wE-bP-gA** (n=10)	
TREE	2.48 ± 0.24	0.54 ± 0.09	3.30 ± 0.30	0.98 ± 0.19
SHRUB	12.69 ± 0.84	2.49 ± 0.12	9.70 ± 0.65	2.40 ± 0.10
FORB	17.38 ± 0.97	3.00 ± 0.15	10.90 ± 0.92	1.86 ± 0.24
GRASS	2.17 ± 0.31	0.70 ± 0.12	2.20 ± 0.39	0.89 ± 0.21
MOSS	1.66 ± 0.28	0.74 ± 0.15	0.00 ± 0.00	0.00 ± 0.00
LICHEN	0.28 ± 0.12	0.10 ± 0.06	0.00 ± 0.00	0.00 ± 0.00
Total	*36.66 ± 2.05*	*3.39 ± 0.09*	*26.10 ± 1.77*	*3.17 ± 0.18*
e2 dogwood bP-wS (n = 21)			**f3 ostrich fern bP-wS** (n = 2)	
TREE	3.10 ± 0.25	1.28 ± 0.09	4.00 ± 0.00	1.79 ± 0.08
SHRUB	10.90 ± 0.86	2.44 ± 0.19	13.00 ± 0.00	2.77 ± 0.51
FORB	16.62 ± 0.96	3.20 ± 0.12	19.00 ± 4.00	2.58 ± 0.63
GRASS	1.67 ± 0.22	0.53 ± 0.14	2.00 ± 1.00	0.75 ± 0.75
MOSS	4.71 ± 0.63	1.49 ± 0.17	4.00 ± 1.00	1.81 ± 0.31
LICHEN	0.90 ± 0.30	0.35 ± 0.15	0.50 ± 0.50	0.00 ± 0.00
Total	*37.90 ± 2.12*	*3.66 ± 0.10*	*42.50 ± 6.50*	*4.14 ± 0.49*

Table 9.   continued

	Richness	Diversity	Richness	Diversity
	g1 Labrador tea–hygric bS-jP (n = 32)		**i1 river alder gully** (n = 6)	
TREE	2.03 ± 0.16	0.47 ± 0.08	0.17 ± 0.17	0.00 ± 0.00
SHRUB	6.22 ± 0.58	1.76 ± 0.16	10.83 ± 1.74	1.58 ± 0.31
FORB	7.31 ± 1.02	2.05 ± 0.23	20.00 ± 1.71	2.84 ± 0.46
GRASS	1.06 ± 0.20	0.39 ± 0.10	4.67 ± 0.49	1.77 ± 0.17
MOSS	4.91 ± 0.34	1.12 ± 0.08	5.17 ± 1.05	2.04 ± 0.43
LICHEN	3.09 ± 0.38	1.19 ± 0.17	0.83 ± 0.40	0.33 ± 0.21
Total	*24.63 ± 1.79*	*2.66 ± 0.10*	*41.67 ± 3.81*	*3.00 ± 0.27*
	h1 horsetail bP-tA (n = 3)		**j1 treed bog** (n = 21)	
TREE	2.00 ± 0.58	0.63 ± 0.33	1.29 ± 0.14	0.07 ± 0.03
SHRUB	11.67 ± 2.60	2.69 ± 0.49	6.19 ± 0.68	1.20 ± 0.14
FORB	17.33 ± 2.60	2.82 ± 0.09	3.38 ± 0.56	0.95 ± 0.17
GRASS	3.67 ± 1.76	1.21 ± 0.83	1.38 ± 0.29	0.43 ± 0.14
MOSS	1.33 ± 0.33	0.32 ± 0.32	7.76 ± 1.04	1.39 ± 0.16
LICHEN	0.33 ± 0.33	0.00 ± 0.00	3.62 ± 0.77	1.27 ± 0.19
Total	*36.33 ± 6.96*	*3.78 ± 0.12*	*23.62 ± 1.92*	*2.77 ± 0.14*
	h2 horsetail bP-wS (n = 10)		**j2 shrubby bog** (n = 6)	
TREE	3.50 ± 0.31	1.52 ± 0.11	0.17 ± 0.17	0.00 ± 0.00
SHRUB	11.30 ± 1.16	2.91 ± 0.18	6.33 ± 0.21	1.45 ± 0.11
FORB	16.70 ± 2.24	2.80 ± 0.19	1.50 ± 0.22	0.34 ± 0.17
GRASS	2.30 ± 0.60	0.84 ± 0.31	0.83 ± 0.40	0.33 ± 0.21
MOSS	3.00 ± 0.39	1.22 ± 0.17	5.33 ± 0.71	0.64 ± 0.17
LICHEN	0.30 ± 0.21	0.10 ± 0.10	6.50 ± 1.26	2.06 ± 0.43
Total	*37.10 ± 3.83*	*3.90 ± 0.10*	*20.67 ± 1.78*	*2.41 ± 0.08*
	h3 horsetail wS-bS (n = 15)		**k1 treed poor fen** (n = 20)	
TREE	2.20 ± 0.31	0.54 ± 0.13	1.65 ± 0.13	0.49 ± 0.10
SHRUB	10.13 ± 0.87	2.74 ± 0.10	7.50 ± 0.44	1.92 ± 0.11
FORB	15.53 ± 2.02	2.30 ± 0.21	7.60 ± 1.14	2.41 ± 0.25
GRASS	2.20 ± 0.31	0.86 ± 0.18	3.30 ± 0.51	1.26 ± 0.17
MOSS	6.00 ± 0.97	1.42 ± 0.22	7.60 ± 0.84	1.57 ± 0.18
LICHEN	0.87 ± 0.29	0.34 ± 0.16	2.10 ± 0.39	0.74 ± 0.18
Total	*36.93 ± 3.61*	*3.39 ± 0.13*	*29.75 ± 2.21*	*3.27 ± 0.10*

Table 9.   concluded

	Richness	Diversity	Richness	Diversity
	k2 shrubby poor fen (n = 9)		**m1 marsh** (n = 15)	
TREE	0.11 ± 0.11	0.00 ± 0.00	0.00 ± 0.00	0.00 ± 0.00
SHRUB	7.33 ± 1.18	1.56 ± 0.26	1.53 ± 0.39	0.40 ± 0.17
FORB	4.78 ± 1.32	1.65 ± 0.40	11.40 ± 1.26	2.23 ± 0.24
GRASS	3.56 ± 0.63	1.30 ± 0.16	6.67 ± 0.97	1.57 ± 0.18
MOSS	5.56 ± 0.75	1.30 ± 0.25	0.93 ± 0.28	0.20 ± 0.14
LICHEN	3.33 ± 1.52	1.13 ± 0.43	0.00 ± 0.00	0.00 ± 0.00
Total	24.67 ± 3.70	2.81 ± 0.15	20.53 ± 2.24	2.58 ± 0.23
	l1 treed rich fen (n = 9)			
TREE	1.44 ± 0.18	0.17 ± 0.08		
SHRUB	8.44 ± 1.04	1.94 ± 0.18		
FORB	8.33 ± 1.70	2.06 ± 0.34		
GRASS	3.44 ± 0.90	0.96 ± 0.30		
MOSS	6.78 ± 1.60	1.55 ± 0.30		
LICHEN	1.11 ± 0.42	0.45 ± 0.23		
Total	29.56 ± 4.50	3.45 ± 0.18		
	l2 shrubby rich fen (n = 22)			
TREE	0.14 ± 0.07	0.00 ± 0.00		
SHRUB	5.05 ± 0.60	0.95 ± 0.12		
FORB	9.32 ± 1.18	2.41 ± 0.17		
GRASS	3.64 ± 0.28	0.97 ± 0.13		
MOSS	4.00 ± 0.60	1.19 ± 0.20		
LICHEN	0.73 ± 0.30	0.25 ± 0.13		
Total	22.86 ± 2.00	2.75 ± 0.10		
	l3 graminoid rich fen (n = 20)			
TREE	0.05 ± 0.05	0.00 ± 0.00		
SHRUB	2.20 ± 0.46	0.86 ± 0.20		
FORB	7.90 ± 1.01	1.97 ± 0.18		
GRASS	4.05 ± 0.46	0.73 ± 0.15		
MOSS	2.00 ± 0.61	0.43 ± 0.17		
LICHEN	0.05 ± 0.05	0.00 ± 0.00		
Total	16.25 ± 1.42	1.79 ± 0.17		

Table 10. Rare and endangered plant species of the **Mid-Boreal ecoregions with biological diversity element ranks**

Latin name	Common name	GRANK	SRANK	Comments
Shrubs				
Arctostaphylos rubra	alpine bearberry	G5	S3	bS/comb. with wB, tL, wS, & shrub; jP
Sorbus scopulina	western mountain-ash	G5	S2	jP-wS
Forbs				
Adoxa moschatellina	moschatel	G5	S3	riparian
Anaphalis margaritacea	pearly everlasting	G5	S2	pine
Andromeda glaucophylla	bog rosemary	G5	S2	bogs, fens
Anemone parviflora	small wood anemone	G5	S1	bS-lichen
Anemone quinquefolia var. *quinquefolia*	wood anemone	G5T?	S?	_ᵃ
Anemone richardsonii	yellow anemone	G5	S1	moist woods, slopes
Antennaria neodioica	broad-leaved pussytoes	G3G5Q	S2	moist grassland, dry open woods, jP
Arethusa bulbosa	swamp-pink	G4	S1	-
Arnica cordifolia	heart-leaved arnica	G5	S3	jP, wS, mixedwood
Aster modestus	great northern aster	G5	S2	moist woods, boggy ground, clearings
Aster pauciflorus	marsh alkali aster	G4	S3	alkaline flats
Aster umbellatus	flat-topped white aster	G5	S2S3	moist woodland, clearings, swampy ground
Bidens frondosa	devil's beggar ticks	G5	S2S3	wet shores, ditches
Calypso bulbosa	Venus's-slipper	G5	S3	-
Campanula aparinoides	marsh bellflower	G5	S2S3	wet sedge fen/marsh
Cardamine pratensis	cuckooflower	G5	S2	sedge swamp, peaty pool, shoreline
Chimaphila umbellata ssp. *occidentalis*	prince's pine	G5T5	S2S3	lP,wS-jP-wB
Cirsium drummondii	Drummond's thistle	G5	S2S3	grassland, open deciduous woods, clearings
Cirsium muticum	swamp thistle	G5	S2	wet open woods, clearings, sloughs
Coeloglossum viride var. *virescens*	long-bracted green orchid	G5	S3	moist meadows, woods
Corallorhiza striata	striped coral-root	G5	S2S3	moist tA

Table 10. continued

Latin name	Common name	GRANK	SRANK	Comments
Forbs continued				
Cypripedium arietinum	ram's-head lady's-slipper	G3	S1	-
Drosera anglica	oblong-leaved sundew	G5	S3	swamps, bogs
Drosera linearis	slender-leaved sundew	G4	S1	bogs
Dryas drummondii	yellow mountain avens	G5?	S1	limestone cliffs
Elatine triandra	longstem waterwort	G5	S2	aquatic
Elodea longivaginata	long-sheath waterweed	G4G5	S2	aquatic
Erigeron hyssopifolius	hyssop-leaved fleabane	G5	S2	shores, banks, ledges, bogs, fens
Erigeron strigosus	daisy fleabane	G5	S2S3	dry open places, prairies, shores, clearings
Geranium carolinianum var. *sphaerospermum*	Carolina crane's-bill	G5T4?	S2S3	open woods, tA - jP
Goodyera oblongifolia	giant rattlesnake-plantain	G5?	S2	wS, Cypress Hills
Lactuca biennis	tall blue lettuce	G5	S2	moist woods, clearings, shrub-thickets
Lilium philadelphicum var. *philadelphicum*	eastern wood lily	G5T?	S1	wB–wS–tA, marshy clearings
Listera borealis	northern twayblade	G5?	S1	wS/salix with carex, boggy bS
Listera cordata	heart-leaved twayblade	G5	S2	wS/feather moss, bS bog
Lomatogonium rotatum	marsh felwort	G5	S2	swamp, bog, meadow, shore, banks
Megalodonta beckii	Beck water-marigold	G4G5	S1	aquatic
Mimulus ringens	square-stem monkeyflower	G5	S1	wet riparian
Myriophyllum alterniflorum	alternate-flowered water milfoil	G5	S1	aquatic
Myriophyllum verticillatum var. *pectinatus*	whorled water milfoil	G5T?	S3	aquatic or with reeds, bul/spike rushes
Najas flexilis	slender naiad	G5	S2	aquatic
Nymphaea tetragona ssp. *leibergii*	white water-lily	G5T5Q	S2	aquatic
Osmorhiza depauperata	blunt-fruited sweet-cicely	G5	S3S4	aspen woods
Parnassia glauca	Carolina grass-of-Parnassus	G5	S2	springs, swamps
Pedicularis groenlandica	elephant's-head	G4G5	S2	sedge meadow, marshes

12-19

Table 10. continued

Latin name	Common name	GRANK	SRANK	Comments
Forbs continued				
Pedicularis macrodonta	swamp lousewort	G4	S2	sedge fen
Pinguicula vulgaris	common butterwort	G5	S2S3	wS feather moss, spring, marl bog
Platanthera dilatata	tall white orchid	G5	S2	moist woods, bog margins, bS/moss
Platanthera orbiculata	round-leaved orchid	G5?	S2S3	coniferous mixedwoods
Polygala paucifolia	fringed milkwort	G5	S2	moist coniferous woods
Polygonum scandens	climbing false buckwheat	G5T?	S1S2	open woods, marsh, shore
Polygonum viviparum	alpine bistort	G5	S1S2	moist woods, meadows
Potentilla multifida	branched cinquefoil	G5	S2	rock outcrops
Prenanthes alba	white rattlesnake-root	G5	S2	deciduous/mixed woods
Rhinanthus minor	yellow rattle	G4	S2	meadows, open woodlands
Saxifraga pennsylvanica	swamp saxifrage	G5	S1	-
Scheuchzeria palustris ssp. *americana*	American scheuchzeria	G5T5	S3	wet fens, open/treed bogs
Senecio pseudaureus	western golden groundsel	G5	S1	meadows, thickets, banks, open woods
Senecio streptanthifolius	northern ragwort	G5	S1	open woods, exposed rocky areas
Spiranthes lacera var. *gracilis*	southern slender ladies'-tresses	G5T4T5	S?	wS-tA, tA-jP, sandy jP, alder
Streptopus amplexifolius var. *amplexifolius*	white twisted-stalk	G5T5	S2S3	wS-bS, wS-tA, wB-wS, springs, seeps
Subularia aquatica var. *americana*	water awlwort	G5T5	S2S3	aquatic
Trientalis europaea ssp. *arctica*	arctic starflower	G4G5T4	S1	moist woods
Trillium cernuum	nodding trillium	G5	S2S3	bF-wS-wB, tA-forb, decid., riparian, wE-mM
Viola conspersa	American bog violet	G5	S1	tA-wB, tA-Salix
Viola pubescens var. *leiocarpon*	yellow violet	G5TU	S1	tA/shrub-fern, riparian, Hudson Bay
Viola selkirkii	great-spurred violet	G5?	S2	wS-Fm, wB-wS, Salix, alder, springs, banks
Viola sororia	wooly blue violet	G5	S1	wS-tA, tA-Salix
Wolffia arrhiza	spotless water-flaxseed	G?	S1	aquatic

12-20

Table 10.   concluded

Latin name	Common name	GRANK	SRANK	Comments
Grass				
Carex pachystachya	sedge	G5	S3	springs, sloughs, tA-jP woods
Carex pedunculata	longstalk sedge	G5	S1	-
Carex petasata	liddon sedge	G5	S2	tA-grass, tA-jP
Carex projecta	necklace sedge	G5	S1	springs, deciduous woods
Carex umbellata	hidden sedge	G5	S?	bS on sand
Carex vulpinoidea	fox sedge	G5	S2	wS-bP-tA woods
Dichanthelium acuminatum	panic grass	G5	S1S2	sand blowout, open woods, clearings, outcrops
Dichanthelium wilcoxianum	Wilcox's panic grass	G5	S1	sandy barrens, prairie, open pine woods
Eleocharis elliptica var. *elliptica*	yellow-seed spike-rush	G5T4?Q	S2	marshes, sandy shores
Eleocharis nitida	slender spike-rush	G3G4	S1S2	-
Juncus nevadensis	sierra rush	G5	S2	slough & stream margins
Juncus stygius ssp. *americanus*	moor rush	G5T5	S1S2	-
Luzula acuminata	hairy wood rush	G5	S1S2	wS-wB-tA-bP-shrub
Luzula multiflora	field wood-rush	G5	S2	sandy woods, rocky shores
Rhynchospora alba	white beak-rush	G5	S2S3	-
Rhynchospora fusca	brown beak-rush	G4G5	S1	-
Scirpus clintonii	Clinton bulrush	G4	S1	open woodland, shores
Scirpus subterminalis	water bulrush	G4G5	S1	aquatic on boggy shores

[a] no comment.

Global Element Ranks [GRANK]: G1 = critically imperiled globally, G2 = imperiled globally, G3 = either very rare and local throughout its range or found locally in a restricted range, G4 = apparently secure globally, G5 = demonstrably secure globally. **Subnational (Provincial) Element Ranks [SRANK]:** S1 = critically imperiled in the province, S2 = imperiled in the province, S3 = rare or uncommon in the province, S4 = widespread, abundant, and apparently secure in the province, but with cause for long-term concern. **OTHER CODES:** ? = inexact or uncertain rank status, Q = questionable taxonomic assignment for current species, T = subrank for species with varieties or subspecies.

system may be used to represent habitats that satisfy some or all of these requirements. The use of a particular ecological unit may vary diurnally and seasonally. Some species require a large area of the same ecological unit and others require a particular combination of ecological units. The ecosystem classification system provides a framework for evaluating, understanding, and communicating information about wildlife, wildlife habitats, and their interaction.

Based largely on information provided by Saskatchewan Environment and Resource Management (SERM), Wildlife Branch[2], the relationships between the ecosystem classification system and the requirements of moose and woodland caribou are discussed. It is anticipated that integrations for other wildlife species will be developed within the classification framework.

Moose

Moose use a variety of habitat types to meet their daily and seasonal requirements for food and cover. Habitats that optimize a mix of these needs are of greatest value. Disturbances such as fire and timber harvesting have the potential to create areas of high quality browse, while areas of mature coniferous and mixedwood forests provide cover. The interface occurring where mature forests meet early successional browse communities are of greatest value to moose. The spatial arrangement of aquatic sites to summer cover and forage, and the juxtaposition of winter cover to browse areas cannot be portrayed easily on the edatope but are important landscape considerations when evaluating moose habitat.

During spring and summer, moose make use of aquatic areas and habitats closely associated with these sites. Calving sites are generally associated with islands and peninsulas, as well as isolated stands of trees in bog (j) and poor fen (k) ecosites and the edges between lowlands and uplands.

Submerged and emergent aquatic vegetation, sedges, grasses, tall forbs, woody browse, and leaves make up the diet of moose during the summer.

---

[2] Genny Greif and Terry Rock, Saskatchewan Environment and Resource Management, Prince Albert, Saskatchewan, personal communications.

Important spring and summer habitats are associated with rich, moist sites where there is an abundance of good browse such as willow (*Salix* spp.) and dogwood (*Cornus stolonifera*) as well as aquatic plant communities. These generally occur on rich fen (l) and marsh (m) ecosites and along many riparian areas. Early successional stages of ecosite phases b2, d2, d3, e1, e2, f1, and f2 provide an abundance of summer forage, while late successional stages (at stand break-up) of hardwood dominant ecosite phases d2, d3, e1, and f1 provide good browse.

During those rare winters where a combination of excessive snow depths and prolonged temperature extremes occur, high density coniferous forests provide areas of less snow accumulation and greater thermal protection. These would typically be provided by ecosite phases b4, d4, e3, and h3 with mature, dense coniferous canopies.

Willows are the primary browse species for moose during the winter. Other browse species include saskatoon, dogwood, aspen, birch (white and dwarf), balsam poplar, pin cherry, and to a lesser extent alder and low-bush cranberry. Similar to summer feeding, these foraging sites are provided by the same early successional and stand break-up stages of ecosite phases b2, d2, d3, e1, e2, f1, and f2. These are of greatest value when situated in close proximity to protective cover.

Woodland Caribou

Woodland caribou are commonly associated with lowland habitats, sedge meadows, and mature to over-mature coniferous forests. There appears to be an affinity for caribou to areas of dwarf birch and sedges with low density, short tamarack along the edges, typically fen (l) and poor fen (k) ecosites, particularly if found within expanses of interconnected lowlands and small lakes. The caribou's summer diet consists of the leaves of dwarf birch and willow, and protein-rich sedges and grasses found in these open lowland areas. These sites are generally poor timber-production areas (wet mineral and organic soils).

During the early winter and late spring caribou forage in poorly drained lowlands of mature black spruce and/or tamarack. Arboreal lichen production is associated with these Labrador tea–hygric (g), bog (j), and poor fen (k) ecosites. In winters of excessive snow accumulation or periods of snow crusting, caribou spend increasingly more time in upland jack pine sites where travel is easier and terrestrial lichens become a dietary staple. Mature and over-mature open-canopied jack pine stands, such as the lichen jP ecosite phase (a1), found on rapidly drained sites with a SV1 soil type are ideal for the production of terrestrial lichens (*Cladina* spp., *Cetraria* spp., and *Cladonia* spp.). High energy terrestrial and arboreal lichens supplemented with sedges are the major component of the caribou's winter diet.

Although woodland caribou do not have extensive migration routes, periodic shifts in range use are important. Caribou tend to follow the same travel routes year after year as they move between areas. The open lowland habitats provide most of the essentials of the caribou's life requirements. Proper management of these lowland habitats and upland jack pine sites, at a landscape scale where spatial distribution patterns are an important consideration, are vital to the continued existence of the caribou.

Clearcutting in jack pine stands is the most effective means of harvesting the timber while ensuring natural pine regeneration. The restoration of terrestrial lichens is also a special consideration on jack pine sites, with winter harvesting having the least impact on ground lichen resources. Large, straight-edged cutblocks with equally large or larger residual (uncut) blocks of similar habitat is probably the best means of removing the timber. Timber harvesting prescriptions that discourage the colonization of areas by white-tailed deer and restrict road access are conducive to good caribou management.

13.0 MENSURATION AND FOREST INVENTORY

The development of this field guide was largely driven by the need for a forest management tool based on ecological function. Because the system is based on function, if applied properly, it has the potential to increase the biological and economic efficiency of silvicultural systems and forest management practices. This section summarizes the forest mensuration data for each ecosite phase, and the site productivity data associated with ecosite phases, soil types, moisture classes, and texture classes. It also includes tables that outline the correlation between the ecological classification system and the provincial forest inventory.

Because of the limited number of plots representing the relationships presented, caution is urged when making interpretations. Please pay attention to the number of samples (n) and the standard error of the mean.

13.1 Forest mensuration

13.1.1 Summary of tree species volume for each forested ecosite phase

Stems per hectare, mean volume per hectare, and mean basal area at breast height per hectare were calculated on a species and plot basis and summarized by ecosite phase (Table 11). Fixed area plots were used to collect the mensuration data. Calculations were based on all live trees greater than 7.0 cm diameter at breast height (DBH). Tree heights that were not measured were estimated from species-specific regression formulas developed from DBH and tree height measurements collected by Saskatchewan Environment and Resource Management (SERM), Forestry Branch, and Geographic Dynamics Corp. Tree volumes were calculated using single-tree volume equations (Northern Systems Centre 1979), based on actual heights, estimated heights, and DBH measurements. Total volume was not standardized to well-stocked levels if stands were over- or under-stocked.

Table 11. Summary of tree species volume for each forested ecosite phase

Tree species	Stems/ha	Mean volume (m³/ha)	Mean basal area (m²/ha)
a1 lichen jP (n = 25)			
black spruce	0.49 ± 0.49	0.00 ± 0.00	0.00 ± 0.00
jack pine	1292.87 ± 134.69	99.46 ± 9.99	19.70 ± 1.47
aspen	2.00 ± 2.00	0.03 ± 0.03	0.01 ± 0.01
white spruce	1.98 ± 1.55	0.03 ± 0.03	0.01 ± 0.01
all species	1297.35 ± 133.94	99.52 ± 9.98	19.73 ± 1.46
b1 blueberry jP-tA (n = 18)			
balsam fir	1.38 ± 0.95	0.04 ± 0.03	0.01 ± 0.01
balsam poplar	0.69 ± 0.69	0.06 ± 0.06	0.01 ± 0.01
black spruce	71.53 ± 41.63	4.44 ± 3.00	0.90 ± 0.58
jack pine	539.35 ± 99.47	140.62 ± 16.54	18.58 ± 2.06
aspen	507.03 ± 111.70	88.86 ± 10.92	11.99 ± 1.50
white birch	28.24 ± 20.24	1.67 ± 1.30	0.33 ± 0.24
white spruce	128.33 ± 68.16	13.10 ± 7.22	2.15 ± 1.11
all species	1276.56 ± 147.07	248.80 ± 13.44	33.97 ± 1.35
b2 blueberry tA(wB) (n = 17)			
balsam fir	8.82 ± 8.08	0.56 ± 0.38	0.11 ± 0.08
black spruce	3.05 ± 2.18	0.45 ± 0.34	0.08 ± 0.05
jack pine	5.88 ± 5.88	0.11 ± 0.11	0.05 ± 0.05
aspen	1087.45 ± 196.80	124.21 ± 17.06	18.41 ± 2.29
white birch	235.16 ± 116.76	20.84 ± 10.42	3.72 ± 1.78
white spruce	42.16 ± 18.80	4.78 ± 2.48	0.82 ± 0.37
all species	1382.52 ± 171.74	150.95 ± 14.29	23.18 ± 1.65
b3 blueberry tA-wS (n = 7)			
balsam poplar	21.43 ± 21.43	2.34 ± 2.34	0.44 ± 0.44
black spruce	3.57 ± 3.57	0.08 ± 0.08	0.03 ± 0.03
jack pine	12.50 ± 10.56	4.09 ± 3.24	0.55 ± 0.43
aspen	512.14 ± 108.56	109.17 ± 24.05	15.18 ± 3.41
white birch	5.34 ± 3.71	0.05 ± 0.04	0.03 ± 0.02
white spruce	383.02 ± 88.56	62.50 ± 18.16	9.74 ± 2.54
all species	937.99 ± 123.20	178.24 ± 27.81	25.96 ± 3.36

Table 11.   continued

Tree species	Stems/ha	Mean volume (m³/ha)	Mean basal area (m²/ha)
b4 blueberry wS(jP) (n = 4)			
balsam fir	3.13 ± 3.13	0.03 ± 0.03	0.01 ± 0.01
black spruce	3.13 ± 3.13	0.94 ± 0.94	0.13 ± 0.13
jack pine	128.13 ± 123.99	41.87 ± 41.57	5.35 ± 5.28
aspen	25.00 ± 17.68	4.58 ± 4.48	0.58 ± 0.54
white spruce	733.33 ± 160.00	165.08 ± 51.40	22.26 ± 6.11
all species	892.71 ± 193.92	212.49 ± 40.76	28.33 ± 5.13
c1 Labrador tea–submesic jP-bS (n = 25)			
black spruce	426.99 ± 178.61	21.57 ± 10.24	4.77 ± 2.10
jack pine	1511.47 ± 183.58	142.45 ± 15.28	25.58 ± 2.07
aspen	4.15 ± 1.90	0.16 ± 0.10	0.04 ± 0.02
white birch	17.79 ± 9.81	1.30 ± 0.81	0.27 ± 0.16
white spruce	3.46 ± 2.08	0.20 ± 0.18	0.04 ± 0.03
all species	1963.86 ± 177.56	165.68 ± 13.59	30.69 ± 1.54
d1 low-bush cranberry jP-bS-tA (n = 4)			
balsam fir	24.86 ± 14.35	1.03 ± 0.84	0.25 ± 0.17
black spruce	6.18 ± 6.18	1.03 ± 1.03	0.18 ± 0.18
jack pine	1559.16 ± 461.92	198.92 ± 31.57	30.32 ± 2.56
aspen	56.01 ± 23.00	21.74 ± 15.05	2.84 ± 1.92
white birch	6.25 ± 3.61	0.42 ± 0.30	0.09 ± 0.06
white spruce	24.93 ± 13.50	5.78 ± 5.04	0.77 ± 0.61
all species	1677.39 ± 416.27	228.93 ± 45.33	34.45 ± 2.19
d2 low-bush cranberry tA (n = 65)			
balsam fir	2.67 ± 2.13	0.15 ± 0.11	0.03 ± 0.02
balsam poplar	76.41 ± 29.63	2.96 ± 0.89	0.75 ± 0.24
black spruce	1.15 ± 0.71	0.08 ± 0.06	0.02 ± 0.01
jack pine	0.77 ± 0.46	0.35 ± 0.27	0.04 ± 0.03
aspen	1288.87 ± 94.63	171.64 ± 11.42	24.43 ± 1.25
white birch	52.55 ± 18.54	4.19 ± 1.62	0.72 ± 0.25
white spruce	46.10 ± 12.65	5.77 ± 2.42	0.92 ± 0.34
all species	1468.51 ± 93.05	185.16 ± 12.31	26.91 ± 1.32

Table 11.   continued

Tree species	Stems/ha		Mean volume (m³/ha)		Mean basal area (m²/ha)	
d3 low-bush cranberry tA-wS (n = 71)						
balsam fir	56.38 ±	16.01	7.63 ±	2.61	1.11 ±	0.35
balsam poplar	49.04 ±	13.18	7.64 ±	2.04	1.22 ±	0.30
black spruce	50.17 ±	22.18	6.02 ±	2.84	1.01 ±	0.47
jack pine	3.51 ±	1.56	1.41 ±	0.62	0.18 ±	0.08
aspen	529.91 ±	61.23	128.70 ±	9.89	16.29 ±	1.14
tamarack	0.18 ±	0.18	0.01 ±	0.01	0.00 ±	0.00
white birch	60.71 ±	18.29	5.74 ±	1.98	0.98 ±	0.30
white spruce	677.06 ±	55.94	126.91 ±	10.61	17.22 ±	1.25
all species	1426.96 ±	78.58	284.07 ±	11.01	38.01 ±	0.99
d4 low-bush cranberry wS (n = 20)						
balsam fir	75.90 ±	36.39	10.50 ±	5.79	1.50 ±	0.80
balsam poplar	29.90 ±	10.82	3.58 ±	1.30	0.67 ±	0.25
black spruce	6.25 ±	3.57	0.97 ±	0.62	0.15 ±	0.10
jack pine	1.24 ±	0.86	0.60 ±	0.47	0.07 ±	0.06
aspen	62.01 ±	21.12	15.14 ±	3.94	2.11 ±	0.59
white birch	36.06 ±	17.02	2.32 ±	1.66	0.46 ±	0.29
white spruce	1334.03 ±	145.47	252.27 ±	17.25	34.21 ±	2.07
all species	1545.38 ±	150.17	285.39 ±	15.85	39.18 ±	1.80
e1 dogwood bP-tA (n = 22)						
balsam fir	11.93 ±	11.35	0.41 ±	0.38	0.10 ±	0.09
balsam poplar	472.73 ±	79.80	156.66 ±	25.73	21.47 ±	3.33
black spruce	2.84 ±	2.32	0.24 ±	0.23	0.04 ±	0.04
aspen	257.28 ±	77.58	69.58 ±	22.26	8.77 ±	2.56
white birch	86.93 ±	52.13	12.49 ±	7.77	1.98 ±	1.22
white spruce	49.43 ±	16.56	4.75 ±	1.92	0.77 ±	0.29
all species	881.14 ±	89.47	244.14 ±	17.10	33.13 ±	1.87

Table 11. continued

Tree species	Stems/ha		Mean volume (m³/ha)	Mean basal area (m²/ha)
e2 dogwood bP-wS (n = 13)				
balsam fir	145.19 ±	78.53	17.76 ± 8.46	2.78 ± 1.31
balsam poplar	177.53 ±	63.46	66.27 ± 19.28	8.96 ± 2.62
black spruce	32.69 ±	23.58	2.95 ± 1.89	0.52 ± 0.34
jack pine	9.62 ±	9.62	3.32 ± 3.32	0.40 ± 0.40
aspen	149.04 ±	53.29	60.87 ± 24.21	7.71 ± 3.02
white birch	105.29 ±	47.93	22.12 ± 11.63	3.25 ± 1.68
white spruce	366.22 ±	102.99	125.19 ± 21.39	15.43 ± 2.88
all species	985.57 ±	122.71	298.49 ± 18.44	39.05 ± 2.00
e3 dogwood wS (n = 8)				
balsam fir	353.12 ±	134.48	51.99 ± 21.36	7.33 ± 2.85
balsam poplar	4.69 ±	4.69	1.05 ± 1.05	0.16 ± 0.16
black spruce	17.19 ±	13.76	2.73 ± 2.08	0.44 ± 0.34
jack pine	7.81 ±	7.81	1.95 ± 1.95	0.25 ± 0.25
aspen	67.19 ±	44.06	16.19 ± 9.45	2.35 ± 1.22
white birch	73.96 ±	49.52	3.60 ± 1.65	0.81 ± 0.41
white spruce	518.23 ±	156.44	223.93 ± 38.80	26.83 ± 4.27
all species	1042.19 ±	169.28	301.44 ± 23.86	38.18 ± 2.14
f1 ostrich fern bP-tA (n = 2)				
balsam poplar	437.50 ±	437.50	158.15 ± 158.15	21.46 ± 21.46
aspen	131.25 ±	131.25	45.55 ± 45.55	6.47 ± 6.47
white birch	118.75 ±	118.75	60.67 ± 60.67	8.22 ± 8.22
all species	687.50 ±	187.50	264.37 ± 51.93	36.16 ± 6.77
f2 ostrich fern mM-wE-bP-gA (n = 9)				
balsam poplar	112.50 ±	53.64	95.66 ± 39.05	13.13 ± 5.20
western cottonwood	1.39 ±	1.39	2.46 ± 2.46	0.42 ± 0.42
green ash	55.56 ±	51.04	8.91 ± 8.83	1.84 ± 1.80
Manitoba maple	554.17 ±	166.13	51.53 ± 13.11	10.92 ± 2.90
white birch	95.83 ±	95.83	17.15 ± 17.15	2.95 ± 2.95
white elm	140.28 ±	64.47	34.87 ± 14.75	5.87 ± 2.48
white spruce	2.78 ±	1.84	3.19 ± 2.98	0.66 ± 0.61
all species	962.50 ±	127.41	213.76 ± 33.04	35.79 ± 4.44

Table 11. continued

Tree species	Stems/ha	Mean volume (m³/ha)	Mean basal area (m²/ha)
f3 ostrich fern bP–wS (n = 2)			
balsam fir	112.50 ± 112.50	25.37 ± 25.37	3.28 ± 3.28
balsam poplar	256.25 ± 68.75	121.74 ± 93.64	19.73 ± 14.12
aspen	18.75 ± 18.75	9.59 ± 9.59	1.59 ± 1.59
white birch	62.50 ± 62.50	4.45 ± 4.45	0.87 ± 0.87
white spruce	175.00 ± 100.00	147.63 ± 83.19	16.02 ± 9.42
all species	625.00 ± 125.00	308.77 ±156.61	41.49 ± 20.97
g1 Labrador tea–hygric bS–jP (n = 26)			
balsam fir	1.28 ± 1.28	0.13 ± 0.13	0.02 ± 0.02
balsam poplar	36.97 ± 18.69	4.13 ± 1.72	0.81 ± 0.34
black spruce	2226.12 ± 249.85	107.00 ± 13.51	23.76 ± 2.36
jack pine	252.70 ± 105.82	23.22 ± 8.81	4.30 ± 1.64
aspen	38.02 ± 17.00	4.94 ± 2.52	0.75 ± 0.35
tamarack	5.89 ± 4.12	0.17 ± 0.12	0.05 ± 0.04
white birch	3.53 ± 2.93	0.32 ± 0.31	0.06 ± 0.06
white spruce	96.39 ± 35.82	16.98 ± 6.59	2.31 ± 0.87
all species	2660.90 ± 210.27	156.89 ± 16.01	32.07 ± 2.07
h1 horsetail bP–tA (n = 1)			
balsam poplar	425.00	123.02	18.60
aspen	675.00	129.33	17.12
white spruce	12.50	2.32	0.35
all species	1112.50	254.68	36.07
h2 horsetail bP–wS (n = 5)			
balsam fir	37.50 ± 37.50	2.90 ± 2.90	0.52 ± 0.52
balsam poplar	187.50 ± 51.69	61.03 ± 15.39	8.64 ± 1.85
black spruce	106.67 ± 77.75	13.57 ± 12.33	2.23 ± 1.92
jack pine	2.50 ± 2.50	0.78 ± 0.78	0.10 ± 0.10
aspen	105.83 ± 53.45	31.59 ± 13.38	4.15 ± 1.55
white birch	12.50 ± 9.68	0.30 ± 0.28	0.09 ± 0.08
white spruce	484.17 ± 146.61	178.33 ± 53.25	20.63 ± 5.36
all species	936.67 ± 196.83	288.51 ± 51.41	36.36 ± 5.13

Table 11.   concluded

Tree species	Stems/ha	Mean volume (m³/ha)	Mean basal area (m²/ha)
h3 horsetail wS-bS (n = 8)			
balsam fir	6.25 ± 6.25	1.62 ± 1.62	0.18 ± 0.18
balsam poplar	46.88 ± 16.83	20.40 ± 9.87	2.57 ± 1.13
black spruce	112.50 ± 77.85	14.80 ± 7.96	2.38 ± 1.32
aspen	14.06 ± 9.86	5.96 ± 3.90	0.96 ± 0.65
tamarack	68.75 ± 68.75	25.48 ± 25.48	3.83 ± 3.83
white birch	6.77 ± 5.22	0.13 ± 0.09	0.04 ± 0.03
white spruce	843.75 ± 145.54	290.93 ± 36.45	34.57 ± 4.11
all species	1098.96 ± 138.67	359.32 ± 32.82	44.54 ± 3.12
j1 treed bog (n = 9)			
black spruce	1624.34 ± 362.36	77.20 ± 16.85	18.05 ± 3.68
tamarack	37.04 ± 20.18	2.30 ± 1.45	0.48 ± 0.27
white spruce	5.56 ± 5.56	0.30 ± 0.30	0.07 ± 0.07
all species	1666.93 ± 359.88	79.80 ± 17.86	18.60 ± 3.81
k1 treed poor fen (n = 9)			
black spruce	464.82 ± 106.84	22.69 ± 8.20	5.16 ± 1.59
tamarack	562.96 ± 132.43	63.61 ± 16.45	11.90 ± 2.89
white birch	1.85 ± 1.85	0.04 ± 0.04	0.01 ± 0.01
white spruce	1.85 ± 1.85	0.05 ± 0.05	0.01 ± 0.01
all species	1031.48 ± 131.54	86.39 ± 17.84	17.09 ± 3.04
l1 treed rich fen (n = 3)			
black spruce	66.67 ± 66.67	2.95 ± 2.95	0.78 ± 0.78
tamarack	1216.67 ± 376.48	91.81 ± 26.00	18.83 ± 1.69
white spruce	8.33 ± 8.33	1.03 ± 1.03	0.17 ± 0.17
all species	1291.67 ± 414.41	95.79 ± 25.73	19.79 ± 1.75

13.1.2 Mean annual increment

Mean annual increment is presented for each forested ecosite phase (Table 12). These values were calculated from estimates of stand volume explained in Section 13.1 and from tree ring analysis. One to three tree cores were taken at breast height from the single dominant species in a plot (generally the species with the greatest volume). Age at breast height was determined from each of the samples and corrected to total tree age using the following age adjustment values[1].

bF	+ 7		tA	+ 4
bP	+ 4		tL	+ 12
bS	+ 20		wB	+ 5
gA	+ 5		wE	+ 5
jP	+ 8		wS	+ 15
mM	+ 5			

Stand age was calculated by averaging total age values derived from age adjusted core readings from the single dominant species. Mean annual increment was calculated by dividing total stand volume (all live trees >7.0 cm DBH) by stand age.

13.1.3 Site index

Site index models for Saskatchewan tree species have been in development over the past several years. These models were applied to one or more species in a plot that were dominant or co-dominant in height. Site index in the context of this field guide is the predicted height of a tree at 50 years breast height age.

Figure 33 depicts site index for each major tree species of each ecosite for the Mid-Boreal ecoregions of Saskatchewan. Table 13 outlines site index for each ecosite phase, Table 14 for each soil type, and Table 15 for each moisture and texture class. Note that n-values in these tables represent the number of plots sampled. Site index was summarized on a plot and species basis before calculating mean values for ecosite phase, soil type, moisture class, and effective texture class.

[1] David Lindenas, Saskatchewan Environment and Resource Management, Prince Albert, Saskatchewan, personal communication.

Table 12. Summary of mean volume, stand age, and mean annual increment for each forested ecosite phase

Ecosite phase	Mean volume (m^3/ha)	Mean age (yrs)	Mean annual increment (m^3/ha/yr)	Number of plots (n)
a1	99.52 ± 9.98	61.8 ± 5.6	1.64 ± 0.16	25
b1	248.80 ± 13.44	83.6 ± 6.0	3.09 ± 0.15	18
b2	163.91 ± 14.71	58.8 ± 3.8	2.85 ± 0.22	13
b3	167.03 ± 29.27	83.8 ± 9.5	2.09 ± 0.37	5
b4	212.49 ± 40.76	108.6 ± 8.6	1.94 ± 0.28	4
c1	168.86 ± 14.45	66.1 ± 5.6	2.67 ± 0.16	23
d1	228.93 ± 45.33	67.8 ± 14.5	3.40 ± 0.21	4
d2	195.91 ± 12.24	57.1 ± 2.7	3.41 ± 0.18	54
d3	290.83 ± 10.93	75.2 ± 2.8	4.03 ± 0.15	67
d4	290.58 ± 15.79	78.8 ± 3.9	3.85 ± 0.27	19
e1	244.60 ± 21.09	71.7 ± 4.5	3.38 ± 0.24	15
e2	313.47 ± 13.44	80.5 ± 5.7	4.04 ± 0.25	11
e3	313.15 ± 24.01	126.0 ± 12.8	2.58 ± 0.22	7
f1	264.37 ± 51.93	n/a[a]	n/a	2
f2	198.62 ± 33.30	71.4 ± 11.1	2.90 ± 0.52	8
f3	465.38	103.0	4.52	1
g1	156.25 ± 16.65	83.4 ± 5.4	2.00 ± 0.22	25
h1	254.68	56.0	4.55	1
h2	333.60 ± 31.90	106.0 ± 7.5	3.15 ± 0.22	4
h3	359.32 ± 32.82	113.4 ± 5.7	3.23 ± 0.32	8
j1	89.67 ± 16.87	121.5 ± 14.3	0.76 ± 0.17	8
k1	85.65 ± 20.21	87.0 ± 14.9	1.02 ± 0.25	8
l1	95.79 ± 25.73	87.3 ± 15.4	1.06 ± 0.10	3

[a] n/a = no stand age data available.
Note: Only those plots with volume and age data have been summarized here.

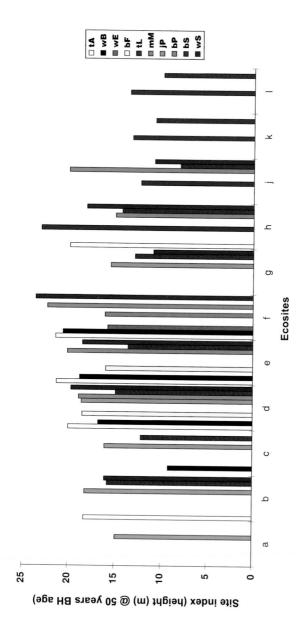

Figure 33.   Site index at 50 years breast height age for the major tree species of each ecosite.

Table 13. Site index at 50 years BH age for each forested ecosite phase

balsam fir (bF)

d3	18.5 ± 1.4 m; n = 4	
e2	15.3 ± 3.0 m; n = 2	
e3	16.5 ± 1.1 m; n = 3	

balsam poplar (bP)

d2	21.1 m; n = 1	
d3	18.4 ± 0.3 m; n = 4	
e1	19.7 ± 0.7 m; n = 12	
e2	21.0 ± 1.4 m; n = 5	
f2	22.7 ± 1.3 m; n = 3	
f3	20.9 m; n = 1	
h1	14.2 m; n = 1	
h2	15.6 ± 1.3 m; n = 2	
h3	14.8 m; n = 1	

black spruce (bS)

b1	15.8 m; n = 1	
c1	12.2 ± 1.3 m; n = 4	
d3	14.9 ± 0.6 m; n = 5	
e1	14.7 m; n = 1	
e3	12.5 m; n = 1	
g1	12.9 ± 0.7 m; n = 23	
h2	14.9 m; n = 1	
h3	14.1 ± 1.3 m; n = 3	
j1	9.0 ± 0.8 m; n = 10	
k1	11.9 ± 1.5 m; n = 5	
l1	9.9 m; n = 1	

jack pine (jP)

a1	14.9 ± 0.3 m; n = 31	
b1	18.2 ± 0.4 m; n = 17	
b4	17.7 m; n = 1	
c1	16.1 ± 0.4 m; n = 25	
d1	19.5 ± 1.2 m; n = 3	
d2	21.5 m; n = 1	
d3	16.7 ± 0.5 m; n = 3	
g1	15.5 ± 1.0 m; n = 6	

Manitoba maple (mM)

f2	16.1 ± 2.1 m; n = 4	

aspen (tA)

b1	18.6 ± 0.4 m; n = 15	
b2	18.8 ± 0.5 m; n = 15	
b3	15.0 ± 0.9 m; n = 4	
b4	19.1 m; n = 1	
d1	20.3 m; n = 1	
d2	20.1 ± 0.3 m; n = 57	

aspen (tA) *(continued)*

d3	20.0 ± 0.4 m; n = 39	
d4	21.0 m; n = 1	
e1	21.0 ± 0.6 m; n = 11	
e2	22.0 ± 1.5 m; n = 4	
e3	22.2 m; n = 1	
f1	20.6 m; n = 1	
f3	22.2 m; n = 1	
h1	20.3 m; n = 1	
h2	21.1 ± 2.6 m; n = 2	
h3	16.9 m; n = 1	

tamarack (tL)

h3	23.0 m; n = 1	
j1	12.3 ± 2.0 m; n = 3	
k1	14.0 ± 1.2 m; n = 9	
k2	6.1 m; n = 1	
l1	13.5 ± 2.0 m; n = 3	

white birch (wB)

c1	9.2 m; n = 1	
d2	18.9 ± 0.2 m; n = 2	
d3	16.0 ± 0.9 m; n = 5	
e1	15.8 m; n = 1	
e2	19.4 ± 1.9 m; n = 5	
f2	20.6 m; n = 1	

white elm (wE)

f2	15.8 ± 0.8 m; n = 3	

white spruce (wS)

b1	18.6 ± 2.5 m; n = 2	
b2	17.3 ± 1.8 m; n = 2	
b3	14.9 ± 1.9 m; n = 3	
b4	14.7 ± 1.7 m; n = 3	
d1	15.4 m; n = 1	
d2	20.9 m; n = 1	
d3	19.8 ± 0.5 m; n = 55	
d4	19.8 ± 0.7 m; n = 22	
e1	19.1 m; n = 1	
e2	19.4 ± 0.5 m; n = 11	
e3	17.0 ± 0.9 m; n = 7	
f2	22.0 m; n = 1	
f3	24.0 ± 2.5 m; n = 4	
g1	10.9 m; n = 1	
h2	18.1 ± 1.4 m; n = 4	
h3	18.0 ± 1.3 m; n = 10	

Table 14. Site index at 50 years BH age for each soil type

balsam fir (bF)

SD4	15.2 m; n = 1
SM3	18.1 m; n = 1
SM4	17.9 ± 1.6 m; n = 5
SWm	17.1 m; n = 1
SR	14.2 m; n = 1

balsam poplar (bP)

SD4	18.2 m; n = 1
SM3	18.8 ± 2.0 m; n = 2
SM4	19.1 ± 0.7 m; n = 18
SMp	22.2 m; n = 1
SWm	21.2 ± 1.0 m; n = 5
SWp	21.2 ± 1.8 m; n = 2
SR	17.4 ± 2.6 m; n = 2

black spruce (bS)

SD1	9.0 m; n = 1
SD4	12.8 ± 0.6 m; n = 3
SM1	16.4 ± 2.2 m; n = 2
SM3	13.5 ± 1.4 m; n = 3
SM4	14.3 ± 0.6 m; n = 8
SMp	15.6 m; n = 1
SWm	13.0 ± 1.1 m; n = 7
SWp	10.8 ± 0.9 m; n = 12
SR	9.2 ± 0.8 m; n = 19

jack pine (jP)

SV1	15.5 ± 0.4 m; n = 27
SV2	14.5 ± 1.6 m; n = 2
SD1	15.0 ± 1.7 m; n = 6
SD2	17.8 ± 0.6 m; n = 2
SD3	21.4 m; n = 1
SD4	16.7 ± 0.5 m; n = 11
SM1	17.0 ± 3.0 m; n = 2
SM2	17.7 ± 0.6 m; n = 2
SM4	17.6 ± 0.6 m; n = 12
SWm	14.6 ± 2.0 m; n = 3

Manitoba maple (mM)

SM4	16.2 ± 3.0 m; n = 3
SWm	15.7 m; n = 1

aspen (tA)

SV1	18.3 ± 0.9 m; n = 10
SV2	19.3 m; n = 1
SD1	19.0 ± 0.5 m; n = 15
SD2	18.6 ± 0.8 m; n = 3
SD3	22.2 ± 0.5 m; n = 3
SD4	18.8 ± 0.6 m; n = 20
SM1	21.0 ± 1.3 m; n = 3
SM2	20.1 m; n = 1
SM3	18.8 ± 1.9 m; n = 3
SM4	20.5 ± 0.3 m; n = 65
SMp	21.9 m; n = 1
SWm	21.2 ± 1.1 m; n = 5
SR	16.9 m; n = 1

tamarack (tL)

SWp	15.4 ± 2.1 m; n = 4
SR	13.2 ± 1.2 m; n = 13

white birch (wB)

SD4	17.9 m; n = 1
SM1	18.4 ± 1.9 m; n = 2
SM4	16.7 ± 1.2 m; n = 10
SMp	13.9 m; n = 1
SWp	24.3 m; n = 1

white elm (wE)

SM3	17.3 m; n = 1
SM4	15.0 ± 0.4 m; n = 2

white spruce (wS)

SV1	17.4 ± 1.8 m; n = 5
SD1	16.9 ± 1.0 m; n = 6
SD2	18.0 ± 1.5 m; n = 3
SD3	22.6 m; n = 1
SD4	17.6 ± 1.1 m; n = 6
SM1	19.0 ± 0.7 m; n = 5
SM2	18.6 ± 1.3 m; n = 3
SM3	19.2 ± 1.9 m; n = 6
SM4	19.7 ± 0.5 m; n = 50
SMp	20.2 ± 5.3 m; n = 2
SWm	19.6 ± 0.4 m; n = 7
SWp	15.4 ± 1.3 m; n = 8
SR	16.8 ± 2.1 m; n = 5

Table 15. Site index at 50 years BH age for each moisture class and effective texture class

Moisture class		Effective texture class	
balsam fir (bF)			
Dry	15.2 m; n = 1	Silty-loamy	18.1 m; n = 1
Moist	17.9 ± 1.3 m; n = 6	Fine loamy-clayey	17.4 ± 1.2 m; n = 7
Wet	15.6 ± 1.4 m; n = 2	Organic	14.2 m; n = 1
balsam poplar (bP)			
Dry	18.2 m; n = 1	Silty-loamy	20.6 ± 2.1 m; n = 3
Moist	19.2 ± 0.6 m; n = 21	Fine loamy-clayey	19.6 ± 0.5 m; n = 26
Wet	20.4 ± 0.9 m; n = 9	Organic	17.4 ± 2.6 m; n = 2
black spruce (bS)			
Dry	11.9 ± 1.0 m; n = 4	Sandy	14.4 ± 1.3 m; n = 6
Moist	14.5 ± 0.6 m; n = 14	Silty-loamy	12.0 ± 1.8 m; n = 4
Wet	10.5 ± 0.6 m; n = 38	Fine loamy-clayey	12.5 ± 0.6 m; n = 26
		Organic	9.2 ± 0.8 m; n = 19
jack pine (jP)			
Very dry	15.4 ± 0.4 m; n = 29	Sandy	15.4 ± 0.4 m; n = 36
Dry	16.3 ± 0.7 m; n = 21	Coarse loamy	16.7 ± 0.8 m; n = 6
Moist	17.5 ± 0.6 m; n = 16	Silty-loamy	21.4 m; n = 1
Wet	14.6 ± 2.0 m; n = 3	Fine loamy-clayey	17.0 ± 0.4 m; n = 25
Manitoba maple (mM)			
Moist	16.2 ± 3.0 m; n = 3	Fine loamy-clayey	16.1 ± 2.1 m; n = 4
Wet	15.7 m; n = 1		
aspen (tA)			
Very dry	18.4 ± 0.8 m; n = 11	Sandy	19.0 ± 0.4 m; n = 29
Dry	19.1 ± 0.4 m; n = 41	Coarse loamy	19.0 ± 0.5 m; n = 5
Moist	20.5 ± 0.3 m; n = 73	Silty-loamy	21.3 ± 1.0 m; n = 8
Wet	20.5 ± 1.1 m; n = 6	Fine loamy-clayey	20.1 ± 0.3 m; n = 88
		Organic	16.9 m; n = 1
tamarack (tL)			
Wet	13.7 ± 1.1 m; n = 17	Sandy	18.0 m; n = 1
		Silty-loamy	13.7 m; n = 1
		Fine loamy-clayey	14.9 ± 4.5 m; n = 2
		Organic	13.2 ± 1.2 m; n = 13
white birch (wB)			
Dry	17.9 m; n = 1	Sandy	16.9 ± 1.9 m; n = 3
Moist	16.8 ± 1.0 m; n = 13	Coarse loamy	24.3 m; n = 1
Wet	24.3 m; n = 1	Fine loamy-clayey	16.8 ± 1.1 m; n = 11
white elm (wE)			
Moist	15.8 ± 0.8 m; n = 3	Silty-loamy	17.3 m; n = 1
		Fine loamy-clayey	15.0 ± 0.4 m; n = 2
white spruce (wS)			
Very dry	17.4 ± 1.8 m; n = 5	Sandy	17.9 ± 0.7 m; n = 17
Dry	17.7 ± 0.7 m; n = 16	Coarse loamy	17.5 ± 1.4 m; n = 8
Moist	19.5 ± 0.4 m; n = 68	Silty-loamy	19.9 ± 1.3 m; n = 11
Wet	17.4 ± 0.8 m; n = 21	Fine loamy-clayey	19.1 ± 0.4 m; n = 66
		Organic	16.8 ± 2.1 m; n = 5

13.2   Forest inventory and site classification

Saskatchewan Environment and Resource Management (SERM), Forestry Branch determines general tree species association, primary tree species, and secondary tree species based on volume (Frey 1981). They have used this approach to determine cover types based on air photo interpretation for the forest inventory maintenance maps (FIMM) and in growth and yield surveys. It is important to determine how volume-derived canopy types in association with drainage classes used in Saskatchewan's FIMM and growth and yield surveys, compare to the ecological classification system.

Since the ecosite phase is defined by forest cover (percent cover) and edatopic position, it is this level of the hierarchical classification system that will be compared with Saskatchewan's FIMM and growth and yield survey system (Frey 1981).

13.2.1   Correlation of volume-derived canopy types with ecosite phases

Using the detailed plot data, the ecosite phase, and the growth and yield survey unit was determined. A brief description about how the detailed plot data were summarized into growth and yield survey units is provided.

Forest mensuration data were used to determine the canopy type for the forested plots. First, volume of all live trees greater than 7.0 cm DBH was calculated to determine the general canopy structure: hardwood (H), softwood (S), or mixedwood (HS, SH). Second, depending on the canopy structure, the dominant hardwood and/or softwood species were assigned to represent the canopy. A moisture regime class (very dry, dry, moist, or wet) was attached to the cover type to represent a complete unit. Correlations were established between these mensuration-derived canopy type/moisture class units and the ecosite phases.

Read down the columns in Table 16 to determine the percent frequency occurrence of each ecosite phase within each canopy type/moisture class unit. Similarly, read down the columns in Table 17 to determine the percent frequency occurrence of each canopy type/moisture class unit within each ecosite phase.

Table 16. Correlation of volume-derived canopy types/moisture class units with each forested ecosite phase. Read down each column to determine the percent frequency occurrence of each forested ecosite phase within each canopy type/moisture class.

	Canopy type/Moisture class								
Ecosite phase	jP Very dry (n = 37)	jP Dry (n = 11)	jP Moist (n = 9)	jP Wet (n = 1)	jPtA Very dry (n = 1)	jPtA Dry-Moist (n = 6)	jPwS Dry (n = 1)	jPbS Dry (n = 1)	jPbS Wet (n = 1)
a1 lichen jP	65								
b1 blueberry jP-tA	5	9	22		100	100	100		
b2 blueberry tA(wB)									
b3 blueberry tA-wS									
b4 blueberry wS(jP)	3								
c1 Labrador tea–submesic jP-bS	27	82	22					100	
d1 low-bush cranberry jP-bS-tA		9	33						
d2 low-bush cranberry tA									
d3 low-bush cranberry tA-wS									
d4 low-bush cranberry wS									
e1 dogwood bP-tA									
e2 dogwood bP-wS									
e3 dogwood wS									
f1 ostrich fern bP-tA									
f2 ostrich fern mM-wE-bP-gA									
f3 ostrich fern bP-wS				100					
g1 Labrador tea–hygric bS-jP			22						100
h1 horsetail bP-tA									
h2 horsetail bP-wS									
h3 horsetail wS-bS									
j1 treed bog									
k1 treed poor fen									
l1 treed rich fen									

Table 16. continued

Ecosite phase		tAjP Very dry (n=2)	tAjP Dry-Moist (n=2)	tA Very dry (n=11)	tA Dry (n=39)	tA Moist (n=42)	tAwB Moist (n=2)	wBtA Very dry (n=1)	wB Dry (n=2)	wBtA Moist (n=1)	wB Moist (n=3)
a1	lichen jP										
b1	blueberry jP-tA	100	100								
b2	blueberry tA(wB)			45	21			100	100		
b3	blueberry tA-wS			9	3	2					
b4	blueberry wS(jP)										
c1	Labrador tea–submesic jP-bS										
d1	low-bush cranberry jP-bS-tA										
d2	low-bush cranberry tA			27	64	69	100				
d3	low-bush cranberry tA-wS			18	8	14					
d4	low-bush cranberry wS										
e1	dogwood bP-tA				3	12					67
e2	dogwood bP-wS					2					33
e3	dogwood wS										
f1	ostrich fern bP-tA									100	
f2	ostrich fern mM-wE-bP-gA										
f3	ostrich fern bP-wS										
g1	Labrador tea–hygric bS-jP										
h1	horsetail bP-tA										
h2	horsetail bP-wS										
h3	horsetail wS-bS										
j1	treed bog										
k1	treed poor fen										
l1	treed rich fen										

Table 16. continued

Ecosite phase	tAwS Very dry (n=1)	tAwS Dry (n=8)	tAwS Moist (n=20)	tAbF Moist (n=1)	wBwS Dry (n=1)	wBwS Moist (n=1)	wBbF Moist (n=1)	wStA Dry (n=4)	wSwB Moist (n=1)
a1 lichen jP	100								
b1 blueberry jP-tA									
b2 blueberry tA(wB)									
b3 blueberry tA-wS		13						50	
b4 blueberry wS(jP)									
c1 Labrador tea-submesic jP-bS									
d1 low-bush cranberry jP-bS-tA									
d2 low-bush-cranberry tA		13	10						
d3 low-bush-cranberry tA-wS		75	85	100	100			50	100
d4 low-bush cranberry wS									
e1 dogwood bP-tA									
e2 dogwood bP-wS			5			100	100		
e3 dogwood wS									
f1 ostrich fern bP-tA									
f2 ostrich fern mM-wE-bP-gA									
f3 ostrich fern bP-wS									
g1 Labrador tea-hygric bS-jP									
h1 horsetail bP-tA									
h2 horsetail bP-wS									
h3 horsetail wS-bS									
j1 treed bog									
k1 treed poor fen									
l1 treed rich fen									

Table 16. continued

Ecosite phase	wStA Moist (n = 20)	wStA Wet (n = 1)	wS Very dry (n = 8)	wS Dry (n = 8)	wS Moist (n = 18)	wSbF Moist (n = 2)	wSbF Wet (n = 1)	wS Wet (n = 11)	bFwS Moist (n = 2)
a1 lichen jP									
b1 blueberry jP-tA									
b2 blueberry tA(wB)									
b3 blueberry tA-wS			13						
b4 blueberry wS(jP)			38						
c1 Labrador tea–submesic jP-bS									
d1 low-bush cranberry jP-bS-tA									
d2 low-bush cranberry tA									
d3 low-bush cranberry tA-wS	85		25	25	28				
d4 low-bush cranberry wS	5			63	61				50
e1 dogwood bP-tA		100							
e2 dogwood bP-wS	5		25	13	6				
e3 dogwood wS					6	50	100	27	50
f1 ostrich fern bP-tA									
f2 ostrich fern mM-wE-bP-gA									
f3 ostrich fern bP-wS						50			
g1 Labrador tea–hygric bS-jP									
h1 horsetail bP-tA	5								
h2 horsetail bP-wS								9	
h3 horsetail wS-bS								64	
j1 treed bog									
k1 treed poor fen									
l1 treed rich fen									

Table 16. continued

Ecosite phase	bF Moist (n = 1)	bFwS Wet (n = 1)	tAbP Moist (n = 3)	bPwB Moist (n = 1)	bPtA Moist (n = 2)	bP Moist (n = 10)	bPtA Wet (n = 1)	bP Wet (n = 7)	bPmM Moist (n = 2)
a1 lichen jP									
b1 blueberry jP-tA									
b2 blueberry tA(wB)									
b3 blueberry tA-wS									
b4 blueberry wS(jP)									
c1 Labrador tea-submesic jP-bS									
d1 low-bush cranberry jP-bS-tA									
d2 low-bush cranberry tA			33		50				
d3 low-bush cranberry tA-wS	100		33	100		10			
d4 low-bush cranberry wS									
e1 dogwood bP-tA					50	70			
e2 dogwood bP-wS		100						86	
e3 dogwood wS									
f1 ostrich fern bP-tA								14	
f2 ostrich fern mM-wE-bP-gA						20			100
f3 ostrich fern bP-wS									
g1 Labrador tea-hygric bS-jP			33						
h1 horsetail bP-tA									
h2 horsetail bP-wS									
h3 horsetail wS-bS							100		
j1 treed bog									
k1 treed poor fen									
l1 treed rich fen									

Table 16. continued

Ecosite phase	wEgA Moist (n = 1)	wEmM Moist (n = 1)	mM Moist-Wet (n = 2)	bPwS Moist (n = 1)	bPwS Wet (n = 3)	wSbP Moist (n = 4)	wSbP Wet (n = 1)	wSbS Moist (n = 1)	wSbS Wet (n = 1)
a1 lichen jP									
b1 blueberry jP-tA									
b2 blueberry tA(wB)									
b3 blueberry tA-wS									
b4 blueberry wS(jP)									
c1 Labrador tea–submesic jP-bS									
d1 low-bush cranberry jP-bS-tA									
d2 low-bush cranberry tA									
d3 low-bush cranberry tA-wS						25			
d4 low-bush cranberry wS									
e1 dogwood bP-tA					100				
e2 dogwood bP-wS						50			
e3 dogwood wS									
f1 ostrich fern bP-tA	100								
f2 ostrich fern mM-wE-bP-gA		100	100						
f3 ostrich fern bP-wS				100					
g1 Labrador tea–hygric bS-jP								100	100
h1 horsetail bP-tA							100		
h2 horsetail bP-wS						25			
h3 horsetail wS-bS									
j1 treed bog									
k1 treed poor fen									
l1 treed rich fen									

Table 16. concluded

Ecosite phase		bSwS Moist (n=1)	bSwS Wet (n=1)	bS Dry (n=2)	bS Moist (n=11)	bS Wet (n=19)	bStL Wet (n=2)	tLwS Wet (n=1)	tLbS Wet (n=4)	tL Wet (n=6)
a1	lichen jP									
b1	blueberry jP-tA									
b2	blueberry tA(wB)									
b3	blueberry tA-wS									
b4	blueberry wS(jP)									
c1	Labrador tea–submesic jP-bS			100						
d1	low-bush cranberry jP-bS-tA				9					
d2	low-bush cranberry tA									
d3	low-bush cranberry tA-wS				9					
d4	low-bush cranberry wS	100								
e1	dogwood bP-tA									
e2	dogwood bP-wS									
e3	dogwood wS									
f1	ostrich fern bP-tA									
f2	ostrich fern mM-wE-bP-gA									
f3	ostrich fern bP-wS		100							
g1	Labrador tea–hygric bS-jP				82	53				
h1	horsetail bP-tA									
h2	horsetail bP-wS									
h3	horsetail wS-bS							100		
j1	treed bog					47				
k1	treed poor fen						100		100	50
l1	treed rich fen									50

Table 17. Correlation of ecosite phases with volume-derived canopy type/moisture class units.
Read down each column to determine the percent frequency occurrence of each canopy type/moisture class unit within each ecosite phase.

Canopy	Moisture class	Ecosite phase										
		a1 (n = 24)	b1 (n = 18)	b2 (n = 17)	b3 (n = 7)	b4 (n = 4)	c1 (n = 25)	d1 (n = 4)	d2 (n = 65)	d3 (n = 71)	d4 (n = 20)	e1 (n = 22)
jP	Very dry	100	11			25	40					
jP	Dry		6				36	25				
jP	Moist		11				8	75				
jPtA	Very dry		6									
jPtA	Dry		22									
jPtA	Moist		11									
jPwS	Dry		6									
jPbS	Dry						4					
tAjP	Very dry		11									
tAjP	Dry		6									
tAjP	Moist		6									
wBtA	Very dry			6								
wB	Dry			12								
wB	Moist											9
tA	Very dry		6	29	14				5	3		
tA	Dry			47	14				38	4		5
tA	Moist			6					45	8		23
tAwB	Moist								3			

Table 17. continued

Canopy	Moisture class	Ecosite phase										
		a1	b1	b2	b3	b4	c1	d1	d2	d3	d4	e1
wBwS	Dry									1		
tAwS	Very dry				14							
tAwS	Dry				14				2	8		
tAwS	Moist								3	24		
tAbF	Moist									1		
wStA	Dry				29					3		
wStA	Moist									24	5	
wSwB	Moist									1		
wS	Very dry				14					3	10	
wS	Dry									3	25	
wS	Moist					75				7	55	
bF	Moist									1		
bFwS	Moist										5	
tAbP	Moist								2			
bPtA	Moist								2			5
bPwB	Moist									1		
bP	Moist								2			32
bP	Wet											27
wSbP	Moist									1		
bSwS	Moist									1		
bS	Dry						8					
bS	Moist						4			1		

Table 17. continued

Canopy	Moisture class	e2 (n = 13)	e3 (n = 8)	f1 (n = 2)	f2 (n = 9)	f3 (n = 2)	g1 (n = 26)	h1 (n = 1)	h2 (n = 5)	h3 (n = 8)	j1 (n = 9)	k1 (n = 9)	l1 (n = 3)
jP	Moist						8						
jP	Wet						4						
jPbS	Wet				11		4						
tA	Moist	8											
wBtA	Moist			50									
wB	Moist												
tAwS	Moist	8											
wBbF	Moist	8											
wBwS	Moist	8											
wStA	Moist	8							20				
wStA	Wet	8											
wS	Dry		13										
wS	Moist	8	13										
wSbF	Moist		13			50							
wSbF	Wet		13										
wS	Wet		38						20				
bFwS	Moist		13							88			
bFwS	Wet	8											

Table 17. concluded

		Ecosite phase											
Canopy	Moisture class	e2	e3	f1	f2	f3	g1	h1	h2	h3	j1	k1	l1
tAbP	Moist							100					
bP	Moist				22								
bPmM	Moist				22								
bPtA	Wet			50					20				
bP	Wet												
wEgA	Moist				11								
wEmM	Moist				11								
mM	Moist				11								
mM	Wet												
bPwS	Moist					50							
bPwS	Wet	23					4						
wSbP	Moist	15							20				
wSbP	Wet								20				
wSbS	Moist						4						
wSbS	Wet						4						
bS	Moist						35						
bS	Wet						38				100		
bStL	Wet									13			
tLwS	Wet											22	
tLbS	Wet											44	
tL	Wet											33	100

13.2.2 Correlation of the forest inventory maintenance maps with ecosite phases

Investigating the relationship between Saskatchewan's forest inventory maintenance maps (FIMM) and the ecosite phase level of the hierarchical ecosystem classification system allows the FIMM to be used as a preliminary classification tool to define ecosite phases.

Each FIMM unit is defined by species association (hardwood, hardwood/softwood, softwood/hardwood, or softwood), canopy height class, crown closure, primary and secondary species, soil drainage, and soil texture. The most meaningful and practical relationship to investigate between the ecosite phases and the FIMM units involves cover type as defined by dominant tree species and soil drainage. The drainage classes that appear on the inventory maps are presented below.

RD	rapidly drained
WD	well drained
MWD	moderately well drained
ID	imperfectly drained
PD	poorly drained

A total of 430 plots were located on the forest inventory maps representing a total of 85 unique map units (map unit = cover type/drainage class). Without losing a significant amount of resolution, the number of unique map units was reduced to 68. In cases where three or more tree species were assigned to a map unit, only the dominant two species were included. For example, a map unit with a tAwBwS canopy was simplified to a tAwS canopy type. Canopies mapped as having a significant understory tree component were grouped with other map units having the same drainage class and the same tree species composition (e.g., tA(wS)/well drained grouped with tAwS/well drained). Map units with the same cover type but with different drainage classes (e.g., jP/well drained versus jP/well to moderately well drained) were not merged to maintain resolution.

The relationships between the ecosite phases and the map units correlate well given the resolution and the range of tree species composition and drainage classes associated with the hierarchical classification system and the FIMM system. There are instances when the correlation is poor. In part, this is the result of the natural variability within a map unit and due to the relatively small size of the sample plot. For example, the gully ecosite (i) is typically too thin to be mapped at the scale of the FIMM and as such tends to be mapped as a forested ecosite. The user should realize that the correlations represent general trends and the relationships should be interpreted and applied cautiously.

Read down the columns in Table 18 to determine the percent frequency occurrence of each ecosite phase within each map unit. Similarly, read down the columns in Table 19 to determine the percent frequency occurrence of map units within each ecosite phase.

Table 18. Correlation of forest inventory maintenance map units with ecosite phases. Read down each column to determine the percent frequency occurrence of each ecosite within each map unit.

		Map units								
Ecosite phase	jP R-W (n = 3)	jP W (n = 15)	jP W-MW (n = 23)	jP MW (n = 3)	jP MW-I (n = 4)	jPbS W (n = 2)	jPbS W-MW (n = 3)	jPbS MW (n = 2)	jPbS MW-I (n = 3)	jPbS I (n = 2)
a1 lichen jP	100	27	43	67		100				
b1 blueberry jP-tA			4	33			33			
b2 blueberry tA(wB)								50		
b3 blueberry tA-wS		7								
b4 blueberry wS(jP)		67								
c1 Labrador tea–submesic jP-bS			39		50		67	50	100	50
d1 low-bush cranberry jP-bS-tA			4		25					
d2 low-bush cranberry tA										
d3 low-bush cranberry tA-wS										
d4 low-bush cranberry wS										
e1 dogwood bP-tA										
e2 dogwood bP-wS										
e3 dogwood wS										
f1 ostrich fern bP-tA										
f2 ostrich fern mM-wE-bP-gA										
f3 ostrich fern bP-wS										
g1 Labrador tea–hygric bS-jP			9		25					50
h1 horsetail bP-tA										
h2 horsetail bP-wS										
h3 horsetail wS-bS										
i1 river alder gully										

Table 18. continued

	Map units									
Ecosite phase	jPtA W-MW (n=3)	jPtA MW (n=2)	jPtA MW-I (n=2)	jPwS MW-I (n=1)	tAjP W (n=3)	tAjP W-MW (n=8)	tAjP MW-I (n=1)	tAwB W (n=2)	tAwB W-MW (n=4)	wBtA W-MW (n=2)
a1 lichen jP										
b1 blueberry jP-tA	67	50	50		33	13	100	50		
b2 blueberry tA(wB)					67	38				50
b3 blueberry tA-wS						13				
b4 blueberry wS(jP)				100						
c1 Labrador tea–submesic jP-bS		50								
d1 low-bush cranberry jP-bS-tA	33		50						50	
d2 low-bush cranberry tA								50	25	
d3 low-bush cranberry tA-wS						13				
d4 low-bush cranberry wS									25	50
e1 dogwood bP-tA						13				
e2 dogwood bP-wS						13				
e3 dogwood wS										
f1 ostrich fern bP-tA										
f2 ostrich fern mM-wE-bP-gA										
f3 ostrich fern bP-wS										
g1 Labrador tea–hygric bS-jP										
h1 horsetail bP-tA										
h2 horsetail bP-wS										
h3 horsetail wS-bS										
i1 river alder gully										

Table 18. continued

Ecosite phase		tA W (n = 10)	tA W-MW (n = 48)	tA MW (n = 9)	tA MW-I (n = 13)	tA I (n = 1)	tAbS W-MW (n = 2)	wBbS W-MW (n = 1)	wBwS MW-I (n = 1)	tAwS W-MW (n = 41)	tAwS MW (n = 5)
a1	lichen jP										
b1	blueberry jP-tA	20	4							5	
b2	blueberry tA(wB)		15	22							
b3	blueberry tA-wS	10					50			5	
b4	blueberry wS(jP)										
c1	Labrador tea–submesic jP-bS										
d1	low-bush cranberry jP-bS-tA	40	63	33	62						
d2	low-bush cranberry tA		2	11	31					7	60
d3	low-bush cranberry tA-wS							100		54	40
d4	low-bush cranberry wS									7	
e1	dogwood bP-tA	20	10	22	8	100				5	
e2	dogwood bP-wS		2				50		100	5	
e3	dogwood wS									5	
f1	ostrich fern bP-tA		4								
f2	ostrich fern mM-wE-bP-gA										
f3	ostrich fern bP-wS										
g1	Labrador tea–hygric bS-jP										
h1	horsetail bP-tA										
h2	horsetail bP-wS									2	
h3	horsetail wS-bS	10									
i1	river alder gully									5	

13-30

Table 18.   continued

Ecosite phase	Map units									
	tAwS MW-I (n = 19)	tAwS I (n = 5)	wStA W-MW (n = 29)	wStA MW (n = 4)	wStA MW-I (n = 14)	wSbF W-MW (n = 3)	wS W-MW (n = 6)	wS MW-I (n = 7)	bPwE MW-I (n = 1)	bP(mM) MW-I (n = 3)
a1 lichen jP										
b1 blueberry jP-tA	5									
b2 blueberry tA(wB)										
b3 blueberry tA-wS			3							
b4 blueberry wS(jP)			3				33			
c1 Labrador tea–submesic jP-bS										
d1 low-bush cranberry jP-bS-tA										
d2 low-bush cranberry tA	11	20								
d3 low-bush cranberry tA-wS	68	20	52	50	86			14		
d4 low-bush cranberry wS	11	60	24		7	33	33	71		
e1 dogwood bP-tA										
e2 dogwood bP-wS			7	25			17			
e3 dogwood wS				25		33	17			
f1 ostrich fern bP-tA										
f2 ostrich fern mM-wE-bP-gA									100	100
f3 ostrich fern bP-wS						33				
g1 Labrador tea–hygric bS-jP										
h1 horsetail bP-tA	5									
h2 horsetail bP-wS			3							
h3 horsetail wS-bS			3		7			14		
i1 river alder gully			3							

Table 18. continued

Ecosite phase	Map units									
	bP(mM) I (n=1)	bP I (n=1)	bPwB I-P (n=1)	bPwS MW-I (n=1)	wSbP W-MW (n=1)	wSbP MW-I (n=2)	wSbS W (n=2)	wSbS W-MW (n=2)	wSbS MW (n=1)	wSbS MW-I (n=3)
b3 blueberry tA-wS										
b4 blueberry wS(jP)										
c1 Labrador tea–submesic jP-bS										
d1 low-bush cranberry jP-bS-tA										
d2 low-bush cranberry tA										
d3 low-bush cranberry tA-wS			100		100					
d4 low-bush cranberry wS								50		33
e1 dogwood bP-tA						100				
e2 dogwood bP-wS										
e3 dogwood wS									100	
f1 ostrich fern bP-tA		100								
f2 ostrich fern mM-wE-bP-gA	100									
f3 ostrich fern bP-wS				100			50			33
g1 Labrador tea–hygric bS-jP										
h1 horsetail bP-tA										
h2 horsetail bP-wS										
h3 horsetail wS-bS								50		33
i1 river alder gully							50			
j1 treed bog										
j2 shrubby bog										
k1 treed poor fen										

Table 18. continued

Ecosite phase	wSbS 1-P (n=1)	bStA W-MW (n=1)	bStA MW-1 (n=1)	bSjP W-MW (n=3)	bSjP MW (n=1)	bSjP MW-1 (n=3)	bSjP 1 (n=1)	bS W-MW (n=4)	bS MW-1 (n=5)	bS 1 (n=4)
b3 blueberry tA-wS								25		
b4 blueberry wS(jP)										
c1 Labrador tea–submesic jP-bS						33				
d1 low-bush cranberry jP-bS-tA			100							
d2 low-bush cranberry tA										
d3 low-bush cranberry tA-wS				33						25
d4 low-bush cranberry wS					100			25	20	
e1 dogwood bP-tA										
e2 dogwood bP-wS										
e3 dogwood wS		100								
f1 ostrich fern bP-tA										
f2 ostrich fern mM-wE-bP-gA										
f3 ostrich fern bP-wS				67						
g1 Labrador tea–hygric bS-jP							100	25	80	75
h1 horsetail bP-tA										
h2 horsetail bP-wS	100									
h3 horsetail wS-bS								25		
i1 river alder gully										
j1 treed bog										
j2 shrubby bog										
k1 treed poor fen						67				

Table 18. concluded

					Map units			
Ecosite phase	bS 1-P (n = 6)	bS P (n = 7)	bStL 1-P (n = 1)	scrub (n = 1)	brushland (n = 11)	treed muskeg (n = 22)	open muskeg (n = 24)	water (n = 9)
d3 low-bush cranberry tA-wS								
d4 low-bush cranberry wS								
e1 dogwood bP-tA								
e2 dogwood bP-wS	17							
e3 dogwood wS		14						
f1 ostrich fern bP-tA								
f2 ostrich fern mM-wE-bP-gA				100				
f3 ostrich fern bP-wS								
g1 Labrador tea–hygric bS-jP	50	57				5		
h1 horsetail bP-tA								
h2 horsetail bP-wS		14						
h3 horsetail wS-bS								
i1 river alder gully					9	5		
j1 treed bog	33	14				14		
j2 shrubby bog					18	14		
k1 treed poor fen			100			23		
k2 shrubby poor fen					9	14	13	
l1 treed rich fen					9	18		
l2 shrubby rich fen					27	9	29	
l3 graminoid rich fen					9		46	11
m1 marsh					18		13	89

Table 19. Correlation of ecosite phases with forest inventory maintenance map units. Read down each column to determine the percent frequency occurrence of each map unit within each ecosite phase.

Canopy	Drainage class	Ecosite phase									
		a1 (n = 22)	b1 (n = 19)	b2 (n = 15)	b3 (n = 5)	b4 (n = 5)	c1 (n = 31)	d1 (n = 4)	d2 (n = 58)	d3 (n = 81)	d4 (n = 22)
jP	R-W	14									
jP	W	18					32				
jP	W-MW	45					29	25			
jP	MW	9	5					25			
jP	MW-I		5				6	25			
jPtA	W-MW		11								
jPtA	MW						3	25			
jPtA	MW-I		5								
jPwS	MW-I					20					
jPbS	W	9									
jPbS	W-MW		5				6				
jPbS	MW		5				3				
jPbS	MW-I			13			10				
jPbS	I						3				
tAjP	W		5								
tAjP	W-MW	5	16	7						1	
tAjP	MW-I		5								
tA	W			13	20				7		
tAwB	W								2		
tAwB	W-MW								3	1	
wBtA	W-MW			7							

Table 19. continued

Canopy	Drainage class	a1	b1	b2	b3	b4	c1	d1	d2	d3	d4
						Ecosite phase					
tA	W-MW		11	47					52	1	
tA	MW			13					5	1	
tA	MW-I								14	5	
tAbS	W-MW				20						
wBbS	W-MW									1	
tAwS	W-MW		11		40				5	27	14
tAwS	MW								5	2	
tAwS	MW-I		5						3	16	
tAwS	I								2	1	
wStA	W-MW					20	3			19	32
wStA	MW									2	
wStA	MW-I									15	
wSbF	W-MW										5
wS	W-MW					40					9
wS	MW-I									1	23
bPwB	I-P										
wSbP	W-MW									1	
wSbS	W-MW									1	5
wSbS	MW-I										5
bStA	MW-I								2		
bSjP	W-MW						3			1	
bSjP	MW-I										
bS	W-MW				20						5
bS	MW-I									5	5
bS	I									1	

Table 19. continued

Canopy	Drainage class	e1 (n = 20)	e2 (n = 16)	e3 (n = 8)	f1 (n = 2)	f2 (n = 8)	f3 (n = 2)	g1 (n = 24)	h1 (n = 1)	h2 (n = 3)	h3 (n = 7)
							Ecosite phase				
jP	W-MW	5						8			
jP	MW-I	10						4			
jPbS	I	5						4			
tAjP	W-MW	5	6								
tA	W	10									
wBtA	W-MW	5									
tAwB	W-MW	5			100						
tA	W-MW	25	6								
tA	MW	10									
tA	MW-I	5									
tA	I	5									
tAbS	W-MW		6								
tAwS	W-MW	10	13	25						33	
wBwS	MW-I		6								
tAwS	MW-I		13						100		
tAwS	I	15									
wStA	W-MW		13	13						33	14
wStA	MW		6								
wStA	MW-I		6	13			50				
wSbF	W-MW			13							
wS	W-MW		6								14

Table 19. continued

Canopy	Drainage class	e1	e2	e3	f1	f2	f3	g1	h1	h2	h3
wS	MW-I										14
bP(mM)	MW-I					38					
bPwE	MW-I					13					
bP	I					13					
bP(mM)	I					13					
bPwS	MW-I					13					
wSbP	MW-I										
wSbS	W							4			
wSbS	W-MW										14
wSbS	MW			13							
wSbS	MW-I							4		33	
wSbS	I-P										14
bStA	W-MW			13							
bSjP	W-MW							8			
bSjP	MW	5									
bSjP	I							4			
bS	W-MW							4			14
bS	MW-I							17			
bS	I							13			
bS	I-P		6					13			
bS	P			13				17			14
scrub	-					13					
treed muskeg	-						50				

Table 19. concluded

		Ecosite phase								
Canopy	Drainage class	i1 (n = 6)	j1 (n = 9)	j2 (n = 5)	k1 (n = 6)	k2 (n = 7)	l1 (n = 5)	l2 (n = 10)	l3 (n = 15)	m1 (n = 14)
tA	W	17								
tA	MW									7
tAwS	W-MW	33								
wStA	W-MW	17								
wSbS	W		11							
bSjP	MW-I		22							
bS	I-P		22							
bS	P		11							
bStL	I-P				17					
scrub	-									
brushland	-	17		40		14	20	30	7	14
treed muskeg	-	17	33	60	83	43	80		13	
open muskeg	-					43		70	73	21
water	-								7	57

14.0 REFERENCES

Achuff, P.L. 1974. Spruce-fir forests of the highlands of northern Alberta. Ph.D. thesis. Department of Botany, Univ. Alberta, Edmonton, Alberta.

Achuff, P.L.; La Roi, G.H. 1977. *Picea-Abies* forests in the highlands of northern Alberta. Vegetatio 33(2/3):127–146.

Agriculture Canada Expert Committee on Soil Survey. 1987. The Canadian system of soil classification. 2nd ed. Agric. Can., Ottawa, Ontario. Publ. 1646.

Alberta Environmental Protection. 1994. Ecological land survey site description manual. Resource Information Branch, Finance, Land Information and Program Support Services, Alberta Environmental Protection.

Alberta Forest Products Association/Alberta Environmental Protection, Land and Forest Services. n.d. Forest soils conservation - Task Force Report. Unpubl. rep.

Atmospheric Environment Service. 1993. Canadian climate normals, 1961–1990; vol 2. Prairie provinces. Environ. Can., Ottawa, Ontario.

Baldwin, K.A.; Sims, R.A. 1989. Field guide to the common forest plants of northwestern Ontario. For. Can., Minist. Nat. Resour., Ontario, Canada-Ontario For. Resour. Development Agreement.

Barbour, M.G.; Burk, J.H.; Pitts, W.D. 1987. Terrestrial plant ecology. Benjamin/Cummings Publishing Company, Inc. Menlo Park, California.

Bauer, D.J. 1976. The climate of Prince Albert National Park. Environ. Can., Atmos. Environ. Serv., Toronto, Ontario. Proj. Rep. 28.

Beckingham, J.D.; Archibald, H. 1996. Field guide to ecosites of northern Alberta. Nat. Resour. Can., Can. For. Serv., Northwest Reg., North. For. Cent., Edmonton, Alberta. Spec. Rep. 5.

Bernier, B. 1968. Descriptive outline of forest humus form classification. Pages 139–154 *in* Proc. 7th Meeting Natl. Soil Survey Comm. Can., Agric. Can., Ottawa, Ontario.

Bloomberg, W. J. 1950. Fire and spruce. For. Chron. 26:157–161.

Boyer, D. 1979. Guidelines for soil resource protection and restoration for timber harvest and post-harvest activities. U.S. Dep. Agric., For. Serv., Pac. Northwest For. Range Exp. Stn., Portland, Oregon.

Braumandl, T.F.; Curran, M.P. (eds.). 1992. A field guide for site identification and interpretation for the Nelson Forest Region. British Columbia Minist. For., Res. Branch, Victoria, British Columbia.

Brierley, D.; Downing, D.; O'Leary, D. 1985. An integrated resource inventory of the Keg River study area. Vol. I and II, Vegetation Classification, Alberta Energy and Natural Resources. Edmonton, Alberta.

Buckman, H.O.; Brady, N.C. 1960. The nature and properties of soils. 6th edition. MacMillan, New York.

Buse, L.J.; Bell, F.W. 1992. Critical silvics of selected crop and competitor species in northwestern Ontario. Ontario Minist. Nat. Resour. Northwestern Ontario For. Tech. Development Unit. Thunder Bay, Ontario.

Carleton, T.J.; Maycock, P.F. 1978. Dynamics of the boreal forest south of James Bay. Can. J. Bot. 56:1157–1173.

Carleton, T.J.; Maycock, P.F. 1981. Understorey - canopy affinities in boreal forest vegetation. Can. J. Bot. 59:1709–1716.

Carr, W. 1982. Surface erosion: hazard assessment and its control. Pages 27–40 *in* Soil interpretations for forestry. British Columbia Minist. For., Victoria, British Columbia. Land Manage. Rep. No.10.

Chee, W.; Vitt, D.H. 1989. The vegetation, surface water chemistry, and peat chemistry of moderate-rich fens in central Alberta, Canada. Wetlands 9(2):227–261.

Clements, F.E. 1928. Plant succession and indicators. H.W. Wilson Co., New York.

Comeau, P.G.; Comeau, M.A.; Utzig, G.F. 1982. A guide to plant indicators of moisture for southeastern British Columbia, with engineering interpretations. British Columbia Minist. For., Victoria, British Columbia.

Corns, I. G. W. 1983. Forest community types of west-central Alberta in relation to selected environmental factors. Can. J. For. Res. 13:995–1010.

Corns, I. G. W.; Annas, R.M. 1986. Field guide to forest ecosystems of west-central Alberta. Can. For. Serv.. North. For. Cent., Edmonton, Alberta.

Cowardin, L.M.; Carter, V.; Golet, F.C.; LaRoe, E.T. 1979. Classification of wetlands and deepwater habitats of the United States. Fish Wildl. Serv., U.S. Dep. Int. Rep. FWS/OBS-79/31.

Daubenmire, R. 1968. Soil moisture in relation to vegetation distribution in the mountains of northern Idaho. Ecology 49:431–438.

Daubenmire, R. 1973. A comparison of approaches to the mapping of forest land for intensive management. For. Chron. 49:87–92.

Daubenmire, R. 1976. The use of vegetation in assessing the productivity of forest lands. Bot. Rev. 42(2):115–143.

DeLong, C.; MacKinnon, A.; Jang, L. 1990. A field guide for identification and interpretation of ecosystems of the northeast portion of the Prince George Forest Region. British Columbia Minist. For., Res. Branch, Victoria, British Columbia.

DeLong, C.; Tanner, D.; Jull, M.J. 1993. A field guide for site identification and interpretation for the southwest portion of the Prince George Forest Region. British Columbia Minist. For., Res. Branch, Victoria, British Columbia.

Dice, L. R. 1945. Measures of the amount of ecologic association between species. Ecology 37:451–460.

Dix, R. L.; Swan, J.M.A. 1971. The roles of disturbance and succession in upland forest at Candle Lake. Can. J. Bot. 49:657–676.

Egan, R.S. 1987. A fifth checklist of the lichen-forming, lichenicolous and allied fungi of the continental United States and Canada. Bryologist 90:77–173.

Ellis, J. G.; Clayton, J.S. 1970. Physiographic divisions of the Northern Provincial Forest. Saskatchewan Inst. Pedol. Publ. SP3.

Ellis, R. A. 1986. Understory development in aspen-white spruce forests in northern Alberta. M.Sc. thesis. Department of Botany. Univ. Alberta, Edmonton, Alberta.

Eulert, G.K.; Hernandez, H. 1980. Synecology and autecology of boreal forest vegetation in the Alberta Oil Sands Environmental Research Program study area. Alberta Oil Sands Environmental Research Program. Edmonton, Alberta. AOSERP Rep. 99.

Farrar, J.L. 1995. Trees in Canada. Fitzhenry and Whiteside Ltd., Markham, Ontario and Nat. Resour. Can., Can. For. Serv., Ottawa, Ontario.

Frey, G. E. 1981. Saskatchewan growth and yield survey field procedures manual. For. Branch, Saskatchewan Nat. Resour. Regina, Saskatchewan.

Froehlich, H.A.; McNabb, D.H. 1983. Minimizing soil compaction in Pacific northwestern forests. Pages 159–192 in E.L. Stone, ed., Forest soils and treatment impacts. Proceedings of the sixth North American Forest Soils Conference, Knoxville, Tennessee. June, 1983. University of Tennessee, Knoxville, Tennessee.

Geographic Dynamics Corp. 1995a. The effects of competition on white spruce growth: analysis of satellite trial B data. Prepared for Canadian Forestry Service, Northern Forestry Centre by Geographic Dynamics Corp., Edmonton, Alberta. Unpubl. rep.

Geographic Dynamics Corp. 1995b. The impacts of forestry practices on boreal mixedwood ecosystems: Edson. Prepared for Canadian Forestry Service, Northern Forestry Centre by Geographic Dynamics Corp., Edmonton, Alberta. Unpubl. rep.

Hall, F. C. 1978. Applicability of rangeland management concepts to forest-range in the Pacific Northwest. Pages 496–499 *in* D.N. Hyder, ed. Proc. First Int. Rangeland Congr., August 14–18, 1978, Denver, Colorado. Soc. Range Manage., Denver, Colorado.

Halliday, W.E.D. 1937. Report on vegetation and site studies, Clear Lake, Riding Mountain National Park, Manitoba. Canada Dep. Int., For. Serv. Res. Note 42.

Harris, W.C. 1980. Guide to forest understory vegetation in Saskatchewan. Saskatchewan Tourism and Renewable Resources, Prince Albert, Saskatchewan. Tech. Bulletin 9.

Hausenbuiller, R.C. 1985. Soil science: principles and practices. 3rd edition. Wm. C. Brown Publishers, Dubuque, Iowa.

Hedrick, D. W. 1975. Grazing mixed conifer clearcuts in northeastern Oregon. Rangeman's Journal 2:6–9.

Heidmann, L.J. 1976. Frost heaving of tree seedlings: a literature review of causes and possible control. U.S. Dep. Agric., For. Serv., Rocky Mt. For. Range Exp. Stn., Fort Collins, Colorado. Tech. Rep. RM-21.

Hitchcock, C.L.; Cronquist, A. 1973. Flora of the Pacific Northwest: an illustrated manual (condensed version). Univ. Washington Press, Seattle, Washington.

Horton, K.W. 1959. Characteristics of subalpine spruce in Alberta. Can. Dep. North. Affairs Nat. Resour., For. Resour. Div. Tech. Note 76.

Ireland, R.R. 1982. Moss flora of the Maritime provinces. National Museums of Canada Publications in Botany, No.13. Ottawa, Ontario.

Ireland, R.R; Brassard, G.R.; Schofield, W.B.; Vitt, D.H. 1987. Checklist of mosses of Canada 2. Lindbergia 13:1–62.

Jeglum, J.K. 1971. Plant indicators of pH and water level in peatlands at Candle Lake, Saskatchewan. Can. J. Bot. 49:1661–1676.

Johnson D.; Kershaw, L.; MacKinnon, A; Pojar, J. (eds.). 1995. Plants of the western boreal forest and aspen parkland. Lone Pine Publishing, Edmonton, Alberta.

Jones, R.K.; Pierpoint, G.; Wickware, G.M.; Jeglum, J.K.; Arnup, R.W.; Bowles, J.M. 1983. Field guide to forest ecosystem classification for the Clay Belt, site region 3E. Agric. Can., Environ. Can. Ontario Minist. Nat. Resour., Toronto, Ontario.

Kabzems, A. 1971. Growth and yield of well stocked white spruce in the mixedwood section Saskatchewan. Dep. Nat. Resour., Province of Saskatchewan, For. Branch. Tech. Bull. 5.

Kabzems, A.; Kosowan, A.L.; Harris, W.C. 1986. Mixedwood section in an ecological perspective Saskatchewan. Saskatchewan Parks and Renewable Resour., For. Div., Tech. Bull. 8.

Klinka K.; Green, R.N.; Trowbridge, R.L.; Lowe, L.E. 1981. Taxonomic classification of humus forms in ecosystems of British Columbia. Land Manage. Rep. 8. British Columbia Minist. For., Victoria, British Columbia.

Knapik, L.J.; Russell, W.B.; Riddell, K.M.; Stevens, N. 1988. Forest ecosystem classification and land system mapping pilot project, Duck Mountain, Manitoba. Prepared for Can. For. Serv. and Man. For. Branch. Unpubl. rep.

Krebs, C.J. 1989. Ecological methodology. Harper and Row Publishers, New York.

Lacate, D.S., compiler. 1969. Guidelines for biophysical land classification. Can. Dep. Fish. For., For. Serv., Ottawa, Ontario. Publ. 1264.

Landon, J.R. 1988. Toward a standard field assessment of soil texture for mineral soils. Soil Survey and Land Evaluation 8:161–165.

Larcher, W. 1983. Physiological plant ecology. 2nd edition. Translated by M.A. Biederman-Thorson. Springer-Verlag, New York.

La Roi, G. H. 1967. Ecological studies in the boreal spruce-fir forests of the North American Taiga. I. Analysis of the Vascular Flora. Ecol. Monogr. 37:229–253.

La Roi, G. H.; Ostafichuk, M. 1982. Structural dynamics of boreal forest ecosystems on three habitat types in the Hondo - Lesser Slave Lake area of north central Alberta. Res. Manage. Div., Alberta Environ. Rec., Parks Wildl. Found. Edmonton, Alberta.

Lee, S.C. 1924. Factors controlling forest successions at Lake Itasca, Minnesota. Bot. Gaz. 78:129–174.

Lindenas, D.G. 1985. Specifications for the interpretation and mapping of aerial photographs in the forest inventory section. Saskatchewan Parks and Renew. Resourc. For. Div., Prince Albert, Saskatchewan. Intern. Doc.

Looman, J.; Best, K.F. 1979. Budd's flora of the Canadian prairie provinces. Res. Branch, Agric. Can. Publ., Quebec, Quebec. Publ. 1662.

Lull, H.W. 1959. Soil compaction on forest and range lands. U.S. Dep. Agric., For. Serv., Washington, D.C., Misc. Publ. 68.

Lutz, H. J. 1956. Ecological effects of forest fires in the interior of Alaska. U.S. Dep. Agric., For. Serv., Alaska For. Res. Cent., Juneau, Alaska. Tech. Bull. 1133.

MacKinnon, A.; Pojar, J.; Coupe, R. (eds.). 1992. Plants of northern British Columbia. Lone Pine Publishing, Edmonton, Alberta.

Martin, N. D. 1959. An analysis of forest succession in Algonquin Park, Ontario. Ecol. Monogr. 29:187–218.

McKee, W.H. Jr.; Hatchell, G.E.; Tiarks, A.E. 1985. Managing site damage from logging. U.S. Dep. Agric., For. Serv., Washington, D.C. Tech. Rep. SE-32.

McNabb, D.H. 1995. Effects of soil modifications on soil physical processes, soil quality, and ecosystem health. Pages 39–58 *in* 32nd Annual Alberta Soil Science Workshop, March 13–15, 1995. Grande Prairie, Alberta.

Meidinger, D.; Pojar, J. (eds.). 1991. Ecosystems of British Columbia. British Columbia Ministry of Forests, Victoria, B.C. Spec. Rep. Ser. 6.

Millar, J.B. 1976. Wetland classification in western Canada. Can. Wildl. Serv., Environ. Can. Saskatoon, Saskatchewan. Rep. Ser. 37.

Monserud, R.A. 1984. Problems with site index: an opinionated review. Pages 167–180 *in* Forest land classification: experiences, problems, perspectives. Symposium proceedings, edited by J. Bockheim, University of Wisconsin, Dep. Soil Sci., Madison, Wisconsin.

Moss, E.H. 1953a. Forest communities in northwestern Alberta. Can. J. Bot. 31:212–252.

Moss, E.H. 1953b. Marsh and bog vegetation in northwestern Alberta. Can. J. Bot. 31:448–470.

Moss, E.H. 1983. Flora of Alberta: a manual of flowering plants, conifers, ferns, and fern allies found growing without cultivation in the province of Alberta, Canada. 2nd Edition. Revised by J.G. Packer. Univ. Toronto Press, Toronto, Ontario.

Mueller-Dombois, D. 1964. The forest habitat types in southeastern Manitoba and their application to forest management. Can. J. Bot. 42:1417–1444.

Mueller-Dombois, D.; Ellenberg, H. 1974. Aims and methods of vegetation ecology. John Wiley & Sons, New York.

National Wetlands Working Group. 1987. The Canadian wetland classification system. Lands Conservation Branch, Can. Wildl. Serv., Environ. Can. Ecol. Land Classif. Ser. 21.

Northern Systems Centre. 1979. 3P sampling system documentation. For. Branch, Saskatchewan Tourism and Renewable Resour., Prince Albert, Saskatchewan. Intern. Doc.

Odum, E.P. 1969. The strategy of ecosystem development. Science 164:262–270.

Ontario Institute of Pedology. 1985. Field manual for describing soils, 3rd edition. Ontario Inst. Pedol. & Univ. Guelph, Ont. OIP Publ. No. 85-3.

Padbury, G.A.; Acton, D.F. 1994. Ecoregions of Saskatchewan 1:2,000,000 scale map. Prepared for Minist. Supply and Serv. Can. and Saskatchewan Property Management Corp. by the Centre for Land and Bio. Resour. Res., Res. Branch, Agriculture and Agri-Food Can., Regina, Saskatchewan.

Pase, C. P. 1958. Herbage production and composition under immature ponderosa pine stands in the Black Hills. J. Range Manage. 11:238–243.

Pojar, J.; Meidinger, D.; Klinka, K. 1991. Concepts. Pages 9–37 *in* Meidinger, D and J. Pojar (eds.) Ecosystems of British Columbia. British Columbia Minist. For., Victoria, British Columbia. Spec. Rep. Ser. 6.

Pritchett, W.L. 1979. Properties and management of forest soils. John Wiley & Sons, New York.

Racey, G.D.; Whitfield, T.S.; Sims, R.A. 1989. Northwestern Ontario forest ecosystem interpretations. Ontario Minist. Nat. Resour. Northwest. Ontario For. Tech. Development Unit, Tech. Rep. 46.

Raup, H. M. 1957. Vegetational adjustment to the instability of the site. Pages 36–48 in: Int. Union, Conserv. Nature Nat. Resour., Proceedings and Papers, 6th Tech. Meeting.

Ross, M.S.; Flanagan, L.B.; La Roi, G.H. 1986. Seasonal and successional changes in light quality and quantity in the understory of boreal forest ecosystems. Can. J. Bot. 64(11):2792–2799.

Rothwell, R.L. 1978. Watershed management guidelines for logging and road construction in Alberta. Environ. Can., North. For. Serv., North. For. Res. Cent., Edmonton, Alberta. Inf. Rep. NOR-X-208.

Rowe, J. S. 1956. Uses of undergrowth plant species in forestry. Ecology 37(3):461–472.

Rowe, J.S. 1959. Forest regions of Canada. Can. Dep. North. Affairs and Nat. Resour. Bulletin 123.

Rowe, J. S. 1961. Critique of some vegetational concepts as applied to forests of northwestern Alberta. Can. J. Bot. 39:1007–1017.

Rowe, J.S. 1972. Forest regions of Canada. Dep. Environ. Can. For. Serv., Ottawa, Ontario. Publ. 1300.

Rowe, J. S., and G. W. Scotter. 1975. Fire in the boreal forest. Quat. Res. 3:444–464.

Sabin, T.E.; Stafford, S.G. 1990. Assessing the need for transformation of response variables. For. Res. Lab., Coll. For., Oregon State Univ., Oregon, U.S. Spec. Publ. 20.

SAS Institute Inc. 1990. SAS/STAT user's guide. SAS Insitute Inc. Version 6, fourth edition, Vol. 1. and 2. Cary, North Carolina.

Schofield, W.B. 1992. Some common mosses of British Columbia. 2nd ed. Roy. B.C. Mus., Handbook 28. Victoria, British Columbia.

Sims, R. A.; Towill, W.D.; Baldwin, K.A.; Wickware, G.M. 1989. Field guide to the forest ecosystem classification for northwestern Ontario. Canada-Ontario Forest Resource Development Agreement.

Singh, P. 1976. Frost damage. Canadian Forest Service, St. John's, Newfoundland. For. Note 13.

Sivak, B. 1991. Field guide to forest ecosystems of southwestern Alberta. Alberta For. Lands Wildl., For. Serv., Edmonton, Alberta.

Slack, N.G.; Vitt, D.H.; Horton, D.G. 1980. Vegetation gradients of minerotrophically rich fens in western Alberta. Can. J. Bot. 58:330–350.

Sousa, W. P. 1984. The role of disturbance in natural communities. Ann. Rev. Ecol. Syst. 15:353–391.

Still, G.; Utzig, G. 1982. Factors affecting the quality of interpretations. Pages 63–73 in: Soil interpretations for forestry. British Columbia Minist. For., Victoria, British Columbia. Land Manage. Rep.10.

Stotler, R.; Crandall-Stotler, B. 1977. A checklist of the liverworts and hornworts of North America. Bryologist 80:405–428.

Strong, W.L. 1994. Pre-harvest ecological assessment handbook. Prepared for Alberta Environmental Protection, Environmental Training Centre by Ecological Land Surveys Ltd., Edmonton, Alberta. Unpubl. rep.

Strong, W.L.; Carnell, B. 1995. Forest site interpretation and silvicultural prescription guide for Alberta. Prepared for Alberta Environmental Protection, Environmental Training Centre by Ecological Land Surveys Ltd., Edmonton, Alberta. Unpubl. rep.

Strong, W. L.; La Roi, G.H. 1983. Rooting depths and successional development of selected boreal forest communities. Can. J. For. Res. 13(4):577–588.

Strong, W.L.; Leggat, K.R. 1991. Ecoregions of Alberta. Prepared for Alberta Forestry, Lands and Wildlife, Land Information Services Division by Ecological Land Surveys Ltd. Edmonton, Alberta. Unpubl. rep.

Swan, J.M.A.; Dix, R.L. 1966. The phytosociological structure of upland forest at Candle Lake, Saskatchewan. J. Ecol. 54:13–40.

Tarnocai, C. 1980. Canadian wetland registry. Pages 9–38 in: C.D.A. Rubec and F.C. Pollett (eds.). Proceedings of a workshop on Canadian wetlands. Lands Directorate, Environ. Can. Ottawa, Ontario. Ecological Land Classification Ser.12.

Ter Braak, C.J.F. 1988. Canoco: a fortran program for canonical community ordination by partial detrended canonical correlation analysis, principle components analysis and redundancy analysis (ver. 2.1). Agric. Math. Group, Wageningen, The Netherlands.

Thorpe, J. 1990. An assessment of Saskatchewan's system of forest site classification. Saskatchewan Res. Counc. Publ. E-2530-1-E-90.

Uresk, D. W.; Severson, K.E. 1989. Understory-overstory relationships in ponderosa pine forests, Black Hills, South Dakota. J. Range Manage. 42(3):203–208.

Viereck, L. A. 1973. Wildfire in the taiga of Alaska. Quat. Res. 3:465–495.

Viro, P. J. 1961. Evolution of site fertility. Unasylva 15:91–97.

Vitt, D.H.; Chee, W. 1990. The relationships of vegetation to surface water chemistry and peat chemistry in fens of Alberta, Canada. Vegetatio 89:87–106.

Vitt, D.H.; Halsey, L.A.; Zoltai, S.C. 1994. The bog landforms of continental western Canada in relation to climate and permafrost patterns. Arctic and Alpine Research 26(1):1–13.

Vitt, D.H.; Marsh, J.E.; Bovey, R.B. 1988. Mosses, lichens, and ferns of northwest North America: a photographic field guide. Lone Pine Publishing, Edmonton, Alberta.

Wali, M.K.; Krajina, V.J. 1973. Vegetation-environment relationships of some sub-boreal spruce zone ecosystems in British Columbia. Vegetatio 26:237–381.

Wamsley, M.; Utzig, G.; Vold, T.; Moon, D.; van Barneveld, J. (eds.). 1980. Describing ecosystems in the field. B.C. Minist. Environ., Resour. Anal. Branch, Victoria, B.C. Tech. Pap.2. and B.C. Minist. For., Res. Branch, Victoria, B.C. Land Manage. Rep.7.

Wischmeier, W.H., and L.D. Meyers. 1973. Soil erodibility on construction areas. U.S. Nat. Res. Counc., Spec. Bulletin 135:20–29.

Working Group on Soil Survey Data. 1983. The Canadian soil information system (CanSIS) manual for describing soils in the field. Res. Branch. Agric. Can.. Ottawa, Ontario.

Zoltai, S.C. 1980. An outline of the wetland regions of Canada. Pages 1–8 in C.D.A. Rubec and F.C. Pollett (eds.). Proceedings of a workshop on Canadian wetlands. Lands Directorate, Environment Canada. Ottawa, Ontario. Ecological Land Classif. Ser. 12.

Zoltai, S.C.; Pollett, F.C. 1983. Wetlands in Canada: their classification, distribution, and use. Pages 245–268 in A.J.P. Gore, ed. Ecosystems of the world. Vol. 4B. Mires: swamp, bog, fen, and moor. Elsevier Scientific Publ., Amsterdam, The Netherlands.

15.0   GLOSSARY OF TERMS

Ah horizon: Dark colored, surface mineral horizon, enriched with humified organic matter (<17% organic carbon by weight).

Aspect: Aspect describes the orientation of a slope as determined by the points of a compass. Aspect in combination with slope, is important in predicting the amount of solar radiation a site receives. A level site has no aspect.

Bog (B): A peatland with weakly to moderately decomposed *Sphagnum* and forest peat material formed in oligotrophic (very poor nutrient status) environments. The bog surface is acidic and low in mineral nutrients due to slightly raised peat surfaces dissociating it from underlying and surrounding mineral-rich soil waters.

Coarse Fragments: The volumetric percentage of rock portions larger than 2 mm in diameter within a soil matrix. Generally they are grouped into three diameter classes: gravels, cobbles, and stones. Fragments from pans or concretions are not considered coarse fragments.

Codominant Plant Species: One of several species that contribute the greatest cover in a plant community type.

Colluvium (C): A mixture of weathered soil and or geological materials transported downslope by gravitational forces and deposited at the base of a slope.

Cover: *See* percent cover.

Decomposition Class: Applicable to organic soils and used to express the degree of organic matter decomposition based on the von Post scale of decomposition (see Appendix 1). There are three main decomposition classes: fibric, mesic, and humic.

Diameter at Breast Height (DBH): The standard diameter measurement of a tree, 1.3 m above the point of germination.

Diversity: Species diversity is a combination of species richness (number of species) and evenness. It is species richness weighted by their relative abundance.

Dominant Plant Species: The species that contributes the greatest cover in a plant community type.

Drainage: Actual water content in excess of field moisture capacity and the extent of the period during which such excess water is present in the plant rooting zone. Soil drainage ranges from very rapid to very poor. Soil drainage and moisture regime are strongly correlated. Drainage class definitions and a key to assist in their evaluation can be found in Section 5.6.6.

Ecoregion: A geographic area that has a distinctive, mature ecosystem on reference sites plus specified edaphic variations as a result of a given regional climate.

Ecosite: Ecological units that develop under similar environmental influences (climate, moisture, and nutrient regime). Ecosites are groups of one or more ecosite phases that occur within the same portion of the edatope (e.g., lichen ecosite). Ecosite, in this classification system, is a functional unit defined by moisture and nutrient regime. It is not tied to specific landforms or plant communities as in other systems (Lacate 1969), but is based on the combined interaction of biophysical factors which together dictate the availability of moisture and nutrients for plant growth. Thus, ecosites are different in their moisture regime and/or nutrient regime.

Ecosite Phase: A subdivision of the ecosite based on the dominant tree species in the canopy. On some sites where a tree canopy is lacking, the tallest structural vegetation layer determines the ecosite phase (i.e., shrubby and graminoid phases). Some variation in humus form or plant species abundance may be observed between ecosite phases. Generally, ecosite phases are mappable units (e.g., a1 lichen jP).

Ecotone: Transition zone between two ecosystems or communities.

Edaphic: Pertains to the soil, particularly with respect to its influence on plant growth and other organisms together with climate.

Edatope: Moisture-nutrient grid that displays the potential ranges of relative moisture (xeric to hydric) and nutrient (very poor to very rich) conditions and outlines relationships between each of the ecosites.

Effective Texture: For mineral soils it is the finest textured soil horizon located 20–60 cm below the mineral soil surface that is thicker than 10 cm. For organic soils it is the dominant decomposition class between 40 and 80 cm below the surface.

Eluviation: The removal of soil material in suspension from a layer or layers of a soil.

Eolian (E): Well sorted, poorly compacted, medium to fine sand and coarse silt sediment that has been transported and deposited by wind. Syn. aeolian.

Ericaceous Plants: A group of low, woody plants belonging to the heath family (Ericaceae).

Feather moss: A collective term for three primary moss species: Schreber's moss (*Pleurozium schreberi*), stair-step moss (*Hylocomium splendens*), and knight's plume moss (*Ptilium crista-castrensis*).

Fen: *See* rich fen and poor fen.

Fibric: Organic material consisting of a large proportion of distinct fiber; poorly decomposed (von Post 1-4) Appendix 1.

Fluvial (F): Moderately well-sorted sediments of gravels and sands with some fractions of silt and clay transported and deposited by streams or rivers.

Fluviolacustrine (FL): Lacustrine deposits that have been partially reworked by fluvial processes.

Forb: Broad-leaved, non-woody plant that dies back to the ground after each growing season (perennial). Ferns and fern allies are considered forbs.

Forb Stratum/Layer: All herbaceous plants exclusive of graminoids as well as some low ericaceous plants that are generally less than 25 cm tall.

Glaciofluvial (GF): Stratified outwash transported and deposited by glacial meltwaters that flowed on, within, under, or beyond the glacier.

Glaciolacustrine (GL): Stratified sediments with generally alternating light and dark bands (varves) deposited in glacial lakes.

Gley: A distinct gray to blue color of mineral soil horizons indicative of long durations of saturation by water. Due to the anaerobic conditions, minerals in these horizons have been reduced and the soil may emit a rotten egg odor (hydrogen sulfide).

Graminoid: Grass-like in form.

Grass Stratum/Layer: Includes all graminoid species (rushes, reeds, sedges, and grasses).

Gully: A narrow but generally deep ravine that usually has flowing water through it for a period of the growing season. The gully may not be easily mapped because of its narrow, linear form.

Hardwood: Stand type that is dominated by deciduous trees such as aspen, white birch, or balsam poplar and has less than 20% total conifer cover in the canopy. It correlates to the forest inventory species association code "H", which is <25% coniferous by volume.

Horizon: A layer of mineral or organic soil, or soil material that has characteristics altered by processes of soil formation.

Humic (Oh): Organic material consisting of only a small proportion of well preserved fiber with most of the material at an advanced stage of decomposition (von Post 7-10) Appendix 1.

Humus Form: Soil horizons located at or near the surface, which have formed from organic residues, either separate from, or intermixed with mineral materials. Humus form can comprise entirely organic or both organic and mineral horizons.

   Moder: Humus form that displays the diagnostic organic horizons with varying degrees of intermixing between the organic and mineral horizons, producing a gradual transition between the horizons.

   Mor: Humus form that displays diagnostic F and H horizons, with a distinct boundary evident between the organic and mineral layer. There is little or no intermixing of organic and mineral horizons.

Mull: Humus form where the diagnostic F and H horizons are commonly lacking. There is considerable mixing of organic material into the surface mineral horizon thereby creating a relatively thick Ah horizon. Usually many soil organisms are present, but it may also form as a result of the decomposition of dense root networks. Insect droppings and earthworms are usually abundant.

Peatymor: Humus form that is strongly associated with lowland, poorly, or very poorly drained soils. It is sharply delineated from the mineral soil and consists of Of, Om, and/or, Oh horizons.

Raw Moder: This humus form is characterized as being transitional between the moders and the fibrimor. It has an L, F, and a thin Hi horizon that is composed of organic granules intermixed with loose mineral grains.

Hydrarch Succession: The directional, cumulative change in plant species composition and abundance on a wet substrate such as in the development of a treed peatland from an open body of water through the gradual accumulation of decomposing organic matter and changing water table levels.

Hydrophyte: Perennial vascular aquatic plants that have their overwintering buds under water.

Illuviation: The translocation or movement of a soil component into a horizon. The process of deposition.

Indicator Species: Plant species that reveal specific site conditions or environmental traits.

Lacustrine (L): Fine sand, silt, and clay sediments deposited on the lake bed or coarser sands that are deposited along a beach by wave action.

Landform: Relief features of the earth's surface produced mainly by erosional and depositional processes.

Lichen: Fungi and certain species of algae that live in a symbiotic relationship whereby the fungus provides structural support, nutrients absorbed from the substrate, and a relatively stable microenvironment. The algae in turn provides carbohydrates through the process of photosynthesis. Generally *Cladina mitis*, *C. arbuscula*, *C. stellaris,* and *C. rangiferina* are considered reindeer lichens.

Lowland: Land that is saturated with water long enough to promote wetland or aquatic processes indicated by poorly drained soils and hydrophytic vegetation.

Marsh (H): A mineral wetland or organic peatland that is periodically inundated up to a depth of 2 m by standing or slowly moving nutrient-rich water. Marshes are typically dominated by emergent rushes, reeds, grasses, and sedges with generally little organic matter accumulation.

Mean Annual Increment (MAI): Total live stem volume of trees that are >7.0 cm diameter at breast height (1.3 m) divided by stand age.

Melanized: A soil horizon that has been darkened and enriched with organic matter (<17% organic carbon by weight).

Mesic (Om): Intermediately decomposed organic matter with properties between those of fibric and humic (von Post 5-6) Appendix 1.

Mixedwood: Stand type that is a blend of deciduous and coniferous trees and has ≥20% and ≤80% total conifer cover in the canopy.

Moder: *See* humus form.

Moisture/Nutrient Grid: *See* edatope.

Moisture Regime (MR): Represents the available moisture supply for plant growth on a relative scale. It is assessed through an integration of species composition and soil and site characteristics. Moisture regime ranges from very xeric to hydric. Syn.=hygrotope. Moisture regime classes and how they are evaluated can be found in Section 5.7.2.

Mor: *See* humus form.

Morainal/Till (M): Sediment generally consisting of well compacted material that is non-stratified and contains a mixture of sand, silt, and clay that has been transported beneath, beside, within, or in front of a glacier.

Moss: A small, leafy plant lacking any true vascular system or roots. Many mosses are indicators of specific environmental conditions.

Mottles: Spots or blotches that contrast with the dominant soil color or matrix color. These are areas of intense oxidation in once saturated (reducing conditions) mineral soils that appear orange to red in color. There are three levels of contrast between mottle color and soil matrix color:

Faint: Evident only on close examination. Faint mottles commonly have the same hue as the color to which they are compared and differ by no more than 1 unit of chroma or 2 units of value. Some faint mottles of similar but low chroma and value can differ by 2.5 units of hue.

Distinct: Readily seen but contrast only moderately with the color to which they are compared. Distinct mottles commonly have the same hue as the color to which they are compared, but differ by 2 to 4 units of chroma or 3 to 4 units of value; or differ from the color to which they are compared by 2.5 units of hue but by no more than 1 unit of chroma or 2 units of value.

Prominent: Contrast strongly with the color to which they are compared. Prominent mottles are commonly the most obvious feature in a soil. Prominent mottles that have medium chroma and value commonly differ from the color to which they are compared by at least 5 units of hue if chroma and value are the same; at least 4 units of value or chroma if the hue is the same; or least 1 unit of chroma or 2 units of value if hue differs by 2.5 units.

MR: *See* moisture regime.

Mull: *See* humus form.

NR: *See* nutrient regime.

Nutrient Regime (NR): Amount of essential nutrients that are available for plant growth. The determination of nutrient regime requires the integration of many environmental and biotic parameters. Soil nutrient regime occurs on a relative scale ranging from very poor to very rich. Syn. trophotope. Nutrient regime classes and information on how they are determined is located in Section 5.7.3.

Organic Matter Thickness: The depth of surface organic horizons in centimetres.

Organic Soil (O): Material of organic origin (plants and animals), in various stages of decay, that has accumulated over time. Generally, the organic matter originates from sedge or *Sphagnum* peat and is subdivided into bog (B), fen (F), swamp (S), marsh (H), or undifferentiated (O) organic components.

Outwash: Material of glaciofluvial origin deposited by meltwater streams, usually in the form of a meltwater channel or an outwash plain.

Parent Material: The surficial material from which soils are formed. It has characteristics that have an important effect on the soil forming process. Classification of parent materials follows that outlined in *The Canadian system of soil classification* (Agriculture Canada Expert Committee on Soil Survey 1987). Parent materials are described in Appendix 4.

Peat: Accumulated remains of mostly dead plants that have decomposed to varying degrees in a wet, usually anaerobic environment; predominantly organic material.

Peat Moss: A synonymous term for sphagnum moss.

Peaty Mor: *See* humus form.

Pedogenesis: The formation of soils.

Percent Cover: An ocular estimation of total ground area that is covered by an individual plant species when its canopy is projected onto the ground surface.

pH: The degree of acidity or alkalinity of soil or groundwater.

Physiography: Pertains to the physical landform characteristics. Syn. geomorphology.

Plant Community Type: A subdivision of the ecosite phase and the lowest taxonomic unit in the classification system. While plant community types of the same ecosite phase share vegetational similarities, they differ in their understory species composition and abundance. These differences may not be mappable from aerial photographs but may be important to wildlife, recreation, and other resource sectors (e.g., a1.2 Pl/blueberry/lichen).

Poor fen: An ecosite that is transitional between the rich fen and bog. A poor fen is intermediate in nutrient regime and is similar floristically to the rich fen and bog. Sedges and peat moss, golden and brown mosses compose the majority of the organic matter content.

Prominence: Plant species occurrence weighted by percent cover

$\sqrt{\% \text{ frequency} \times \% \text{ cover}}$.

Raw Moder: *See* humus form.

Reference Site: A site that is more strongly influenced by the regional climate than by edaphic or landscape factors. Generally, this site has well to moderately well drained soils with medium textures that neither lack nor have an excess of soil nutrients or moisture.

Reindeer Lichen: An important plant in the woodland caribou diet that is composed of an algae and a fungus growing together to their mutual benefit (symbiosis). Generally *Cladina mitis*, *C. arbuscula*, *C. stellaris,* and *C. rangiferina* are considered reindeer lichens.

Residual (X): Unconsolidated or partly weathered bedrock as the parent material of a soil, presumed to have developed in place from the consolidated rock on which it lies.

Rich Fen (N): A peatland with moderate to well decomposed sedge, grass, and reed peat material formed in eutrophic environments. Mineral-rich waters are at or just above the rich fen surface. *Sphagnum* is usually absent or subordinate to other mosses.

Richness: Species richness is the number of species in a given area or community.

Seepage: The subsurface, gravitational movement of water through a site.

Seral: Series of stages that follow one another in an ecological succession. A seral stage is one stage in the successional sequence.

Shrub Stratum/Layer: Woody perennials of smaller structure than trees occupying the stratum from ground level to 5 m in height. Tree species can occur in this stratum during younger developmental stages.

SI: *See* site index.

Site Index (SI): Predicted height for a specific tree species at a given breast height age (50 years in this guide).

Slope: The slope of a site describes the percentage of vertical rise relative to horizontal distance. Zero degrees as percent slope describes a level site and 45° is equivalent to 100% slope.

Softwood: Stand type that is dominated by coniferous trees such as spruce, fir, pine, and tamarack and has greater than 80% total conifer cover in the canopy. It correlates to the forest inventory species association code "S", which is >75% coniferous by volume.

Soil: The naturally occurring unconsolidated mineral or organic material at least 10 cm thick that occurs at the earth's surface and is capable of supporting plant growth.

Soil Horizon: A layer of mineral or organic soil or soil material that has been altered by pedogenic or soil forming processes.

Soil Moisture Class: Classes used in the soil type designation based on moisture regime (MR) values including the following codes: very dry (V), dry (D), moist (M), and wet (W).

Soil Particle Size: Relative proportions of the different sized soil particles. The particle size distribution of the soil's fine fraction (less than 2 mm) determines the soil texture.

Soil Perviousness: The potential of a soil to allow water to pass through it. It is affected by soil texture, structure, porosity, cracks, and shrink-swell properties.

Soil Textural Class: Classes of soil texture used in the soil type designation include the following codes: sandy (1), coarse loamy (2), silty-loamy (3), and fine loamy-clayey (4).

Soil Texture: The relative proportion of sand, silt, and clay in a mineral soil.

Soil Type: Soil types are functional taxonomic units used to stratify soils based on soil moisture regime, effective soil texture, organic matter thickness, and solum depth. Soil types can be used in association with the hierarchical classification system (ecosite, ecosite phase, and plant community type) and to classify disturbed sites.

Soil Type Modifier: An "open legend" used in association with the soil type to provide more resolution. They are soil attributes that significantly influence tree productivity or are considered to have important implications for management. They include organic matter thickness, humus form, surface coarse fragment content, and surface texture.

Succession: The change in community composition over time following a major disturbance. This change is measured by the composition of the plant populations, by their reproduction and distribution on the substrate, and by their coexistence. Generally, ecosite phases within an ecosite are closely related successional stages. The end stage of succession is referred to as climax (see seral).

Surface Texture: For mineral soils it is the dominant soil texture within the top 20 cm. For organic soils it is the dominant decomposition class within the top 40 cm.

Swamp (S): Wooded mineral wetland or peatland with standing water or water gently flowing through pools or channels that persist for long periods.

Symbiosis: Intimate association of two species, usually involving coevolution.

Textural Triangle: A graphical representation showing the relationship between 14 different particle size distributions of mineral soil materials. The proportion of sand and clay are represented along the horizontal and vertical axes respectively.

Till: *See* morainal.

Topographic Position: The relative location of a site on the landscape. Positions range from crest to depression. Topographic positions are described in Section 5.7.1.

Tree Stratum/Layer: Trees in this layer occur in the uppermost canopy or in the understory, providing they are greater than 5 m tall. Only commercial tree species are included, i.e., not willow or alder.

Varve: Alternating dark and light colored bands of sediment formed in pro-glacial lakes by glacial meltwaters. One pair of bands is equivalent to one year of sediment accumulation.

Vegetation Strata: Layers of plant growth based on morphology and normal height of all species. These layers are used on the fact sheets to describe the appearance of the dominant vegetation components of the ecosite phase and plant community type. Primary strata include: Tree, Shrub, Forb, Grass, Moss, and Lichen.

von Post Scale of Organic Matter Decomposition: Refer to decomposition class and Appendix 1.

Water Table: The uppermost level of water in a zone of saturated soil that has a long duration of saturation. Not to be confused with seepage.

APPENDIX 1: VON POST SCALE OF ORGANIC MATTER DECOMPOSITION

In this field test, first squeeze a sample of organic material within a closed hand to remove excess water. Then squeeze the sample a final time and observe: 1) the distinctness of the plant structure in the material both before and after it is squeezed; 2) the color of the solution that is expressed from the sample during squeezing; 3) the proportion of the original sample that extrudes between the fingers during squeezing; and 4) the nature of the residual material in hand. Ten classes are defined.

Fibric [Of]

1. Undecomposed

Plant structure unaltered; yields only clear, light yellow brown colored water.

2. Almost undecomposed

Plant structure distinct; yields only clear, light, yellow brown colored water.

3. Very weakly decomposed

Plant structure distinct; yields distinctly turbid brown water; no peat substance passes between the fingers; residue not mushy.

4. Weakly decomposed

Plant structure distinct; yields strongly turbid brown water; no peat substance escapes between fingers; residue rather mushy.

Mesic [Om]

5. Moderately decomposed

Plant structure clear but becoming indistinct; yields much turbid brown water; some peat escapes between the fingers; residue very mushy.

6. Strongly decomposed

Plant structure somewhat indistinct but clearer in the squeezed residue than in the undisturbed peat; yields much turbid brown water; about a third of the peat escapes between the fingers; residue strongly mushy.

Humic [Oh]

7. Strongly decomposed

Plant structure indistinct but recognizable; about half of the peat escapes between the fingers.

8. Very strongly decomposed

Plant structure very indistinct; about two thirds of the peat escapes between the fingers; residue almost entirely of resistant remnants such as root fibers and wood.

9. Almost completely decomposed

Plant structure almost unrecognizable; nearly all of the peat escapes between the fingers.

10. Completely decomposed

Plant structure unrecognizable; all of the peat escapes between the fingers.

Note: This appendix is based on Agriculture Canada Expert Committee on Soil Survey (1987).

APPENDIX 2: SOIL HORIZON DESIGNATIONS

Definitions follow the conventions of *The Canadian system of soil classification* (Agriculture Canada Expert Committee on Soil Survey 1987).

Organic Soil Horizons

LFH Organic horizons developed primarily from the accumulation of leaves, twigs, or woody materials with or without a minor component of mosses (but not including the living moss layer of the forest floor); usually not saturated with water for prolonged periods; containing >17% organic carbon (approximately 30% organic matter) by weight.

L Organic horizon characterized by the accumulation of mainly leaves (and needles), twigs, and woody materials; most of the original structures are easily discernible.

F Organic horizon characterized by an accumulation of partly decomposed organic matter derived mainly from leaves, twigs, and woody materials; some of the original structures are easy to recognize; the material may be somewhat altered by soil fauna as in a moder, or it may be a partly decomposed mat permeated by fungal hyphae as in a mor.

H Organic horizon is characterized by an accumulation of decomposed organic matter in which the original structures are indiscernible; differing from the F horizon by having greater humification chiefly due to the action of soil organisms; it may be sharply delineated from the mineral soil as in a mor, where humification is chiefly dependent on fungal activity, or it may be (partly) incorporated into the mineral soil as in moders and mulls (see Hi horizon).

Hi Organic horizon characterized by an accumulation of spherical or cylindrical organic granules (animal droppings); intermixing with mineral particles is common; genetically it is an intermediate stage between an H and an Ah horizon.

Of,Om,Oh   Organic horizons developed mainly from the accumulation of mosses, rushes, and woody material; containing >17% organic carbon (approximately 30% organic matter) by weight.

Of (fibric)   Organic horizon consisting of fibric materials (von Post 1–4); containing large amounts of distinct fiber.

Om (mesic)   An intermediately decomposed organic horizon (von Post 5–6) with properties between those of fibric and humic materials.

Oh (humic)   Organic horizon consisting of humic materials (von Post 7–10); containing only small amounts of well preserved fiber with most of the substrate at an advanced stage in decomposition.

Mineral Soil Horizons

A   Mineral A horizon formed at or near the surface in the zone of leaching or eluviation of materials in solution or suspension, maximum in situ accumulation of organic matter or both.

Ae   Mineral A horizon characterized by the eluviation of clay, minerals, or organic matter alone or in combination. When dry it is usually lighter in color than underlying B horizons.

Ah   Mineral A horizon enriched with organic matter; containing <17% organic carbon (approximately 30% organic matter) by weight. An Ah horizon is usually darker than the underlying mineral horizons. Ah horizons may be buried by lighter colored mineral materials (Ahb). Buried horizons commonly occur in floodplain areas where alluvial materials are deposited.

B Mineral B horizon characterized by the enrichment of organic matter, sesquioxides, or clay, by the development of soil structure or by a change of color denoting hydrolysis, reduction, or oxidation.

Bg Mineral B horizon that has gray-blue colors, mottling, or both indicative of permanent or periodic intense reduction. Less intense periods of reduction may cause faint or distinct mottling (Bmgj, Btgj).

Bt Mineral B horizon that contains illuvial clay. It forms below an eluvial horizon, such as the Ae horizon. It usually has a higher ratio of fine clay to total clay than the unaltered parent material of the C horizon.

Bm Mineral B horizon that has been slightly altered by hydrolysis, oxidation, or solution to give a change in color, structure, or both.

C Mineral C horizon comparatively unaffected by any pedogenic processes or weathering observed in A or B horizons, except the process of gleying (Cg) and the accumulation of additional carbonates (Cca). Generally, the C horizon can be used to describe parent material.

Ck Mineral C horizon containing carbonates as indicated by visible effervescence when dilute HCl (10% by volume) is added to the soil matrix.

APPENDIX 3: GREAT GROUPS AND SUBGROUPS OF THE CANADIAN SYSTEM OF SOIL CLASSIFICATION

Brunisolic order

E.DYB	Eluviated Dystric Brunisol
E.EB	Eluviated Eutric Brunisol
GLE.DYB	Gleyed Eluviated Dystric Brunisol
GLE.EB	Gleyed Eluviated Eutric Brunisol
GL.EB	Gleyed Eutric Brunisol
O.EB	Orthic Eutric Brunisol

Cryosolic order

FI.OC	Fibric Organic Cryosol
GL.SC	Gleyed Static Cryosol

Gleysolic order

HU.LG	Humic Luvic Gleysol
O.G	Orthic Gleysol
O.HG	Orthic Humic Gleysol
O.LG	Orthic Luvic Gleysol
R.G	Rego Gleysol
R.HG	Rego Humic Gleysol

Luvisolic order

BR.GL	Brunisolic Gray Luvisol
D.GL	Dark Gray Luvisol
GLBR.GL	Gleyed Brunisolic Gray Luvisol
GLD.GL	Gleyed Dark Gray Luvisol
GL.GL	Gleyed Gray Luvisol
O.GL	Orthic Gray Luvisol

Organic order

CU.M	Cumulic Mesisol
FI.M	Fibric Mesisol
HY.F	Hydric Fibrisol
T.F	Terric Fibrisol
T.H	Terric Humisol
T.M	Terric Mesisol
TME.F	Terric Mesic Fibrisol
TY.F	Typic Fibrisol
TY.M	Typic Mesisol

Regosolic order

CU.R	Cumulic Regosol
GLCU.R	Gleyed Cumulic Regosol
GL.HR	Gleyed Humic Regosol
GL.R	Gleyed Regosol
O.HR	Orthic Humic Regosol

Note: This appendix is based on Agriculture Canada Expert Committee on Soil Survey (1987). Only those soils that appear on the fact sheets are listed.

APPENDIX 4: PARENT MATERIAL DEFINITIONS

Mineral

Colluvium (C)

Colluvial materials are formed in response to the downslope movement of materials due to gravity. This process is often referred to as mass wasting. They are usually found in lower slope positions due to the nature of their deposition (gravity). They may be found along any slope, especially where the concavity or convexity changes. The texture of these deposits is dependent on the material from which they originated. Due to the gently sloping nature of most of the study area, colluvial deposits are not common. Examples of colluvial landforms are slumps and slope failures.

Eolian (E)

Eolian deposits consist of materials deposited by wind. These postglacial deposits are often well sorted and primarily consist of medium to fine sands. They do not normally show evidence of banding or differential deposition. Sand dunes are a common example of this type of parent material. In areas bordering dune fields, a thin layer of eolian sand is often present over other materials such as moraine or lacustrine deposits. Eolian deposits are often found in association with glaciofluvial deposits. Silty textured eolian deposits (loess) are sometimes found on terraces adjacent to major rivers.

Fluvial (F)

Fluvial parent materials and landforms are developed through the action of flowing water (i.e., rivers and streams). The types of materials and landforms associated with these geomorphic features are highly variable. Common examples of fluvial features include river terraces, banks between terraces and along modern streams, meander scars and oxbows (i.e., abandoned portions of a river channel), and alluvial fans (i.e., fan-shaped deposits from a stream where an abrupt decrease in the stream gradient occurs).

Fluvioeolian (FE)

Sediments that have been deposited or reworked by fluvial and eolian processes which may or may not have been active at the same time. The deposits cannot be discretely separated as fluvial or eolian.

Fluviolacustrine (FL)

Fluviolacustrine materials are lacustrine deposits that have been partially reworked by fluvial processes. Typically, fluviolacustrine deposits are sands or loamy sands with thin bands of dark colored, medium to moderately fine-textured sediments. Topography ranges from gently undulating to strongly rolling but generally most landscapes are knoll and depression type.

Glaciofluvial (GF)

Glaciofluvial materials were deposited by glacial meltwaters (temporary glacial creeks and rivers) and are generally coarse-textured sands and gravels. They may be thick blankets or thin veneers covering glaciolacustrine or morainal materials. This stratified drift can vary in its water-holding capacity over short distances due to textural changes in the materials that were deposited by varying flow velocities of the glacial meltwater. Gravelly deposits occur where glacial meltwater velocities were high while sands occur where the water velocities were slower. In some instances, the deposits are slightly finer in texture near the surface as a result of decreased flow velocities during the final stages of deposition.

The topography of the bedrock of the interior plains and the existing morainal landforms influenced the flow path of the meltwater streams that deposited the glaciofluvial materials. This may have been the case where the glacial meltwater was flowing under and beyond the glacier. When the meltwater was within or on top of the glacier, however, the glaciofluvial materials could have been deposited in any topographic position.

Glaciolacustrine (GL)

Glaciolacustrine sediments were deposited in standing water. The retreating ice fronts formed a large number of extensive, short-lived preglacial lakes, which resulted in large areas of the boreal forest being capped with glaciolacustrine deposits. Supraglacial lakes were prevalent on the retreating ice sheet and contributed to the extensive distribution of lacustrine materials. These materials are usually fine-textured (i.e., very fine sand to clays) and sorted into relatively uniform particle sizes that often show evidence of banding. Near the shoreline of a lake, bands of sand often occur between bands of finer deposits. Within the shoreline zone, beach deposits and wave-cut slopes may be present. The topography is often a relatively uniform plain or long slope, if the deposits are thick enough to mask the underlying terrain.

Lacustrine (L)

Lacustrine materials are postglacial deposits generally consisting of either stratified fine sand, silt, or clay deposited on the lake bed. Moderately well sorted and stratified sand and coarser materials are transported and deposited by wave action. Lacustrine deposits, like glaciolacustrine deposits, are typically found on relatively uniform plains or long slopes, if the deposits are thick enough to mask the underlying terrain.

Moraine/Till (M)

Till consists of materials that have been transported by glacial ice. These materials are deposited in landforms known as moraines. Such deposits are often poorly sorted (i.e., consist of a mixture of particle sizes ranging from clay to boulders) and unstratified (i.e., not consisting of distinctive zones of similarly sized materials). Topographically, morainal deposits often have an irregular surface that ranges from undulating to hummocky, but also occur as plains and distinctive ridges. Topography ranges from level to strongly sloping.

Rock (R)

Consolidated component (bedrock) comprising materials that are tightly packed or indurated. This includes igneous, metamorphic, sedimentary, and consolidated volcanic rocks (bedrock).

Organic

Bog (B)

Sphagnum or forest peat materials formed under an ombrotrophic (nutrient poor) environment due to the slightly elevated nature of the bog tending to be dissociated from nutrient-rich groundwater or surrounding mineral soils.

Fen (N)

Sedge peat materials derived primarily from sedges with inclusions of partially decayed woody material formed in a eutrophic (nutrient rich) environment due to the close association of the material with mineral-rich waters.

Marsh (H)

Deposits composed of mineral or organic material with a high mineral content, but with little peat accumulation.

Swamp (S)

A peat-covered or peat-filled area with the water table at or above the peat surface. The dominant peat materials are shallow to deep mesic to humic forest and fen peat formed in a eutrophic (nutrient rich) environment resulting from strong water movement from the margins or other mineral sources.

Undifferentiated Organic (O)

Deposits with any of the criteria for bog, fen, swamp, or marsh, which have not been differentiated.

Note: Only those parent materials that appear on the fact sheets are listed.

APPENDIX 5: SOIL ASSOCIATIONS

Soil association	Map unit	Parent material
Dominantly Brunisolic and Regosolic soils		
Kewanoke	Kk	Coarse-textured, gravelly, weakly to non-calcareous, glaciofluvial and glaciolacustrine sands, some of which have been reworked by the wind
Pine	Pn	Coarse-textured, weakly to non-calcareous, glaciofluvial and glaciolacustrine sands, some of which have been reworked by the wind
Sipanok	Sk	Medium to fine-textured, strongly calcareous, stratified, recent levee deposits; common along the Cumberland Delta of the Mid-Boreal Lowlands; Cumulic and Gleyed Cumulic Regosols are common
Dominantly Luvisolic soils		
Bittern Lake	Bt	Medium to moderately fine-textured, moderately calcareous glacial till, overlain by moderately coarse to medium-textured materials
Bodmin	Bd	Coarse to moderately coarse-textured, gravelly glaciofluvial deposits
Bow River	Bo	Moderately coarse to coarse-textured, weakly calcareous, sandy glacial till
Dorintosh	Do	Moderately fine to fine-textured, moderately calcareous glaciolacustrine deposits

Soil association	Map unit	Parent material
Dominantly Luvisolic soils (continued)		
Flotten	Ft	Coarse to moderately coarse-textured glaciofluvial and glaciolacustrine deposits, overlying moderately fine to fine-textured glaciolacustrine deposits
Kakwa	Kw	Fine-textured, moderately to strongly calcareous sandy glaciolacustrine deposits
La Corne	Lc	Moderately coarse to medium-textured, weakly to moderately calcareous sandy, glaciolacustrine deposits containing greater than 15% clay
Loon River	Ln	Medium to moderately fine-textured, weakly to moderately calcareous glacial till
Piprell	Pr	Medium to moderately fine-textured, moderately calcareous glacial till, overlain by coarse to moderately coarse-textured stony materials
Porcupine Plain	Pp	Medium to moderately fine-textured, moderately to strong calcareous, silty glaciolacustrine deposits
Waitville	Wv	Medium to moderately fine-textured banded glaciofluvial and glaciolacustrine deposits
Dominantly Gleysolic soils		
Arbow	Aw	Variable-textured, glacial and recent deposits of poorly drained depressional areas, black spruce vegetation
Marsh	Mh	Very poorly drained, marshy margins of lakes and estuaries, rush and sedge vegetation

Soil association	Map unit	Parent material
Dominantly Gleysolic soils (continued)		
Meadow	Mw	Variable-textured, glacial and recent deposits of poorly drained depressional areas, meadow vegetation
Miscellaneous mineral soils		
Alluvium	Av	Variable-textured, recent fluvial deposits associated with river floodplains; Regosolic and Gleysolic soils are common
Beach	Bx	Variable-textured, sandy and gravelly deposits that are often stony, occurring as beach ridges and shore lines
Hillwash	Hw	Variable-textured, colluvial deposits of valley slopes and eroded escarpments
Organic soils		
Bowl Bog	Bb	Fibric moss peat overlying mesic to humic forest peats, 60–120 cm thick
Flat Bog	Bf	Fibric moss peat overlying mesic peat of sedge, moss, or forest region, 60–180 cm or more thick
Stream Bog	Bs	Mesic to fibric moss and forest peat, 60–120 cm thick
Bowl Fen	Fb	Mesic sedge peats 40–120 cm thick
Horizontal Fen	Fh	Open, sedge-dominated peatlands occupying extensive depressional and lowland areas
Patterned Fen	Fp	Mesic sedge peat commonly greater than 120 cm thick
Stream Fen	Fs	Mesic to humic sedge peat 40–180 cm or more thick

APPENDIX 6: SOIL SURVEY REFERENCES

Anderson, D.W.; Ellis, J.G. 1976. The soils of the provincial forest reserves in the Prince Albert Map Area (73-H Sask.). Sask. Inst. Pedol., Univ. Sask., Saskatoon, Saskatchewan. Publ. SF3.

Ayres, K.W.; Anderson, D.W.; Ellis, J.G. 1978. The soils of the northern provincial forest in the Pasquia Hills and Saskatchewan portion of The Pas Map Area (63-E and 63-F Sask.). Sask. Inst. Pedol., Univ. Sask., Saskatoon, Saskatchewan. Publ. SF4.

Crosson, L.S.; Ellis, J.G.; Shields, J.A. 1970. The soils of the northern provincial forest reserves in the Shellbrook Map Sheet (73-G Sask.). Sask. Inst. Pedol., Univ. Sask., Saskatoon, Saskatchewan. Publ. SF.1.

Head, W.K.; Anderson, D.W.; Ellis, J.G. 1981. The soils of the Wapawekka Map Area (73-I Sask.). Sask. Inst. Pedol., Univ. Sask., Saskatoon, Saskatchewan. Publ. SF5.

Padbury, G.A. 1984. The soils of the provincial forests in the Green Lake-Waterhen Map Areas (73J-K), Sask. Inst. Pedol., Univ. Sask., Saskatoon, Saskatchewan. Publ. SF6. Maps only.

Rostad, H.P.W.; Ellis, J.G. 1972. The soils of the provincial forest in the St. Walburg Map Area (73-F Sask.). Sask. Inst. Pedol., Univ. Sask., Saskatoon, Saskatchewan. Publ. SF2.

Saskatchewan Institute of Pedology, 1983. The soils of the provincial forests in Amisk-Cormorant Lake Map Areas (63L-K) . Sask. Inst. Pedol., Univ. Sask., Saskatoon, Saskatchewan. Maps only.

Saskatchewan Soil Survey. 1995. The soils of Cote rural municipality no. 271. Sask. Inst. Pedol., Univ. Sask., Saskatoon, Saskatchewan.

Saskatchewan Soil Survey. 1995. The soils of St. Philips rural municipality no. 301. Sask. Inst. Pedol., Univ. Sask., Saskatoon, Saskatchewan.

Saskatchewan Soil Survey. 1995. The soils of rural municipality no. 331. Sask. Inst. Pedol., Univ. Sask., Saskatoon, Saskatchewan.

Saskatchewan Soil Survey. 1995. The soils of rural municipality no. 333. Sask. Inst. Pedol., Univ. Sask., Saskatoon, Saskatchewan.

Saskatchewan Soil Survey. 1995. The soils of rural municipality no. 334. Sask. Inst. Pedol., Univ. Sask., Saskatoon, Saskatchewan.

Saskatchewan Soil Survey. 1995. The soils of rural municipality no. 335. Sask. Inst. Pedol., Univ. Sask., Saskatoon, Saskatchewan.

Saskatchewan Soil Survey. 1995. The soils of rural municipality no. 394. Sask. Inst. Pedol., Univ. Sask., Saskatoon, Saskatchewan.

Saskatchewan Soil Survey. 1995. The soils of rural municipality no. 395. Sask. Inst. Pedol., Univ. Sask., Saskatoon, Saskatchewan.

Saskatchewan Soil Survey. 1995. The soils of rural municipality no. 456. Sask. Inst. Pedol., Univ. Sask., Saskatoon, Saskatchewan.

Saskatchewan Soil Survey. 1995. The soils of rural municipality no. 464. Sask. Inst. Pedol., Univ. Sask., Saskatoon, Saskatchewan.

Saskatchewan Soil Survey. 1995. The soils of rural municipality no. 465. Sask. Inst. Pedol., Univ. Sask., Saskatoon, Saskatchewan.

Saskatchewan Soil Survey. 1995. The soils of rural municipality no. 467. Sask. Inst. Pedol., Univ. Sask., Saskatoon, Saskatchewan.

Saskatchewan Soil Survey. 1995. The soils of rural municipality no. 486. Sask. Inst. Pedol., Univ. Sask., Saskatoon, Saskatchewan.

Saskatchewan Soil Survey. 1995. The soils of rural municipality no. 487. Sask. Inst. Pedol., Univ. Sask., Saskatoon, Saskatchewan.

Saskatchewan Soil Survey. 1995. The soils of rural municipality no. 488. Sask. Inst. Pedol., Univ. Sask., Saskatoon, Saskatchewan.

Saskatchewan Soil Survey. 1995. The soils of rural municipality no. 490. Sask. Inst. Pedol., Univ. Sask., Saskatoon, Saskatchewan.

Saskatchewan Soil Survey. 1995. The soils of rural municipality no. 491. Sask. Inst. Pedol., Univ. Sask., Saskatoon, Saskatchewan.

Saskatchewan Soil Survey. 1995. The soils of rural municipality no. 494. Sask. Inst. Pedol., Univ. Sask., Saskatoon, Saskatchewan.

Saskatchewan Soil Survey. 1995. The soils of rural municipality no. 496. Sask. Inst. Pedol., Univ. Sask., Saskatoon, Saskatchewan.

Saskatchewan Soil Survey. 1995. The soils of rural municipality no. 497. Sask. Inst. Pedol., Univ. Sask., Saskatoon, Saskatchewan.

Saskatchewan Soil Survey. 1995. The soils of rural municipality no. 498. Sask. Inst. Pedol., Univ. Sask., Saskatoon, Saskatchewan.

Saskatchewan Soil Survey. 1995. The soils of rural municipality no. 499. Sask. Inst. Pedol., Univ. Sask., Saskatoon, Saskatchewan.

Saskatchewan Soil Survey. 1995. The soils of rural municipality no. 501. Sask. Inst. Pedol., Univ. Sask., Saskatoon, Saskatchewan.

Saskatchewan Soil Survey. 1995. The soils of rural municipality no. 502. Sask. Inst. Pedol., Univ. Sask., Saskatoon, Saskatchewan.

Saskatchewan Soil Survey. 1995. The soils of rural municipality no. 520. Sask. Inst. Pedol., Univ. Sask., Saskatoon, Saskatchewan.

Saskatchewan Soil Survey. 1995. The soils of rural municipality no. 521. Sask. Inst. Pedol., Univ. Sask., Saskatoon, Saskatchewan.

Saskatchewan Soil Survey. 1995. The soils of rural municipality no. 555. Sask. Inst. Pedol., Univ. Sask., Saskatoon, Saskatchewan.

Saskatchewan Soil Survey. 1995. The soils of Loon Lake rural municipality no. 561. Sask. Inst. Pedol., Univ. Sask., Saskatoon, Saskatchewan.

Saskatchewan Soil Survey. 1995. The soils of rural municipality no. 595. Sask. Inst. Pedol., Univ. Sask., Saskatoon, Saskatchewan.

Saskatchewan Soil Survey. 1995. The soils of Beaver River rural municipality no. 622. Sask. Inst. Pedol., Univ. Sask., Saskatoon, Saskatchewan.

Stonehouse, H.B.; Ellis, J.G. 1983. The soils of the Hudson Bay and Saskatchewan portion of the Swan Lake Map Areas (63-D and 63-C Sask.). Sask. Inst. Pedol., Univ. Sask., Saskatoon, Saskatchewan. Publ. S5.

APPENDIX 7: PLANT NAMES

Abies balsamea (L.) Mill. .... balsam fir

Acer negundo L. .... Manitoba maple

Acer spicatum Lamb. .... mountain maple

Achillea millefolium L. .... common yarrow

Actaea rubra (Ait.) Willd. .... red and white baneberry

alder-leaved buckthorn .... ***Rhamnus alnifolia*** L'Hér.

Alnus crispa (Ait.) Pursh .... green alder

Alnus tenuifolia Nutt. .... river alder

Amelanchier alnifolia Nutt. .... saskatoon

Andromeda polifolia L. .... bog rosemary

Apocynum androsaemifolium L. .... spreading dogbane

Aralia nudicaulis L. .... wild sarsaparilla

Arctostaphylos uva-ursi (L.) Spreng. .... bearberry

aspen .... ***Populus tremuloides*** Michx.

Aster ciliolatus Lindl. .... Lindley's aster

Aster conspicuus Lindl. .... showy aster

Athyrium filix-femina (L.) Roth .... lady fern

Aulacomnium palustre (Hedw.) Schwaegr. .... tufted moss

awned hair-cap .... ***Polytrichum piliferum*** Hedw.

balsam fir .... ***Abies balsamea*** (L.) Mill.

balsam poplar .... ***Populus balsamifera*** L.

bastard toad-flax .... ***Geocaulon lividum*** (Richards.) Fern.

beaked hazelnut .... ***Corylus cornuta*** Marsh.

beaked willow .... ***Salix bebbiana*** Sarg.

bearberry .... *Arctostaphylos uva-ursi* (L.) Spreng.

beautiful sedge .... *Carex concinna* R.Br.

Betula papyrifera Marsh. .... white birch

Betula pumila L., *Betula glandulosa* Michx. .... dwarf birch

bishop's-cap .... *Mitella nuda* L.

black spruce .... *Picea mariana* (Mill.) B.S.P.

blueberry .... *Vaccinium myrtilloides* Michx.

bog cranberry .... *Vaccinium vitis-idaea* L.

bog rosemary .... *Andromeda polifolia* L.

Brachythecium spp. .... ragged moss

bracted honeysuckle .... *Lonicera involucrata* (Richards.) Banks

bristly black currant .... *Ribes lacustre* (Pers.) Poir.

Bromus ciliatus L. .... fringed brome

brown moss .... *Drepanocladus* spp.

buck-bean .... *Menyanthes trifoliata* L.

bulrush .... *Scirpus* spp.

bunchberry .... *Cornus canadensis* L.

bush honeysuckle .... *Diervilla lonicera* Mill.

Calamagrostis canadensis (Michx.) Beauv. .... marsh reed grass

Calamagrostis stricta (Timm) Koeler .... narrow reed grass

Caltha palustris L. .... marsh marigold

Canada buffalo-berry .... *Shepherdia canadensis* (L.) Nutt.

Canada goldenrod .... *Solidago canadensis* L.

Canada thistle .... *Cirsium arvense* (L.) Scop.

Carex concinna R.Br. .... beautiful sedge

Carex spp. .... sedge

cattail .... *Typha latifolia* L.

Ceratodon purpureus (Hedw.) Brid. .... fire moss

Chamaedaphne calyculata (L.) Moench .... leatherleaf

choke cherry .... ***Prunus virginiana*** L.

Cinna latifolia (Trev.) Griesb. .... drooping wood-reed

Circaea alpina L. .... small enchanter's nightshade

Cirsium arvense (L.) Scop. .... Canada thistle

Cladina mitis (Sandst.) Hale & W. Culb. .... reindeer lichen

Cladina spp. .... reindeer lichen

Cladonia borealis S. Stenroos .... red pixie-cup

Cladonia deformis (L.) Hoffm. .... deformed cup lichen

Climacium dendroides (Hedw.) Web. & Mohr .... common tree moss

cloudberry .... ***Rubus chamaemorus*** L.

common beaked moss .... ***Eurhynchium pulchellum*** (Hedw.) Jenn.

common great bulrush ....***Scirpus validus*** Vahl

common hair-cap .... ***Polytrichum commune*** Hedw.

common horsetail .... ***Equisetum arvense*** L.

common pink wintergreen .... ***Pyrola asarifolia*** Michx.

common tree moss .... ***Climacium dendroides*** (Hedw.) Web. & Mohr

common nettle .... ***Urtica dioica*** L.

common yarrow .... ***Achillea millefolium*** L.

Cornus canadensis L. .... bunchberry

Cornus stolonifera Michx. .... dogwood

Corylus cornuta Marsh. .... beaked hazelnut

cream-colored vetchling .... ***Lathyrus ochroleucus*** Hook.

creeping spike-rush .... ***Eleocharis palustris*** (L.) R. & S.

currant .... ***Ribes*** spp.

cushion moss ....***Dicranum polysetum*** Sw.

cushion moss .... ***Dicranum undulatum*** Brid.

deformed cup lichen .... *Cladonia deformis* (L.) Hoffm.

dewberry .... *Rubus pubescens* Raf.

Dicranum polysetum Sw. .... cushion moss

Dicranum undulatum Brid. .... cushion moss

Diervilla lonicera Mill. .... bush honeysuckle

dogwood .... *Cornus stolonifera* Michx.

Drepanocladus spp. .... brown moss

Drepanocladus uncinatus (Hedw.) Warnst. .... sickle moss (brown moss)

drooping wood-reed .... *Cinna latifolia* (Trev.) Griesb.

Dryopteris carthusiana (Vill.) H.P. Fuchs .... shield fern

dwarf birch .... *Betula pumila* L., *Betula glandulosa* Michx.

dwarf scouring-rush .... *Equisetum scirpoides* Michx.

Eleocharis palustris (L.) R. & S. .... creeping spike-rush

Elymus innovatus Beal .... hairy wild rye

Epilobium angustifolium L. .... fireweed

Epilobium ciliatum Raf. .... northern willowherb

Equisetum arvense L. .... common horsetail

Equisetum pratense Ehrh. .... meadow horsetail

Equisetum scirpoides Michx. .... dwarf scouring-rush

Equisetum spp. .... horsetail

Equisetum sylvaticum L. .... woodland horsetail

Eurhynchium pulchellum (Hedw.) Jenn. .... common beaked moss

false melic .... *Schizachne purpurascens* (Torr.) Swallen

fire moss .... *Ceratodon purpureus* (Hedw.) Brid.

fireweed .... *Epilobium angustifolium* L.

fowl bluegrass .... *Poa palustris* L.

Fragaria vesca L. .... woodland strawberry

Fragaria virginiana Duchesne .... wild strawberry

Fraxinus pennsylvanica Marsh. .... green ash

fringed brome .... *Bromus ciliatus* L.

Galium boreale L. .... northern bedstraw

Galium triflorum Michx. .... sweet-scented bedstraw

Geocaulon lividum (Richards.) Fern. .... bastard toad-flax

giant reed grass .... *Phragmites australis* (Cav.) Trin. ex Steud.

golden moss .... *Tomenthypnum nitens* (Hedw.) Loeske

great-spurred violet .... *Viola selkirkii* Pursh

green alder .... *Alnus crispa* (Ait.) Pursh

green ash .... *Fraxinus pennsylvanica* Marsh.

Gymnocarpium dryopteris (L.) Newm. .... oak fern

hairy wild rye .... *Elymus innovatus* Beal

high-bush cranberry .... *Viburnum opulus* L.

horsetail .... *Equisetum* spp.

Hudsonia tomentosa Nutt. .... sand heather

Hylocomium splendens (Hedw.) B.S.G. .... stair-step moss

jack pine .... *Pinus banksiana* Lamb.

Juncus balticus Willd. .... wire rush

Juncus spp. .... rush

Kalmia polifolia Wang. .... northern laurel

kidney-leaved violet .... *Viola renifolia* A. Gray

knight's plume moss .... *Ptilium crista-castrensis* (Hedw.) De Not.

Labrador tea .... ***Ledum groenlandicum*** Oeder

lady fern .... ***Athyrium filix-femina*** (L.) Roth

Larix laricina (Du Roi) K. Koch .... tamarack

late goldenrod .... ***Solidago gigantea*** Ait.

Lathyrus ochroleucus Hook. .... cream-colored vetchling

leafy moss .... ***Plagiomnium*** spp.

leatherleaf .... ***Chamaedaphne calyculata*** (L.) Moench

Ledum groenlandicum Oeder .... Labrador tea

Lindley's aster .... ***Aster ciliolatus*** Lindl.

Linnaea borealis L. .... twin-flower

Lonicera dioica L. .... twining honeysuckle

Lonicera involucrata (Richards.) Banks .... bracted honeysuckle

low-bush cranberry .... ***Viburnum edule*** (Michx.) Raf.

Lycopodium annotinum L. .... stiff club-moss

Maianthemum canadense Desf. .... wild lily-of-the-valley

Manitoba maple .... ***Acer negundo*** L.

marsh cinquefoil .... ***Potentilla palustris*** (L.) Scop.

marsh marigold .... ***Caltha palustris*** L.

marsh reed grass .... ***Calamagrostis canadensis*** (Michx.) Beauv.

marsh skullcap .... ***Scutellaria galericulata*** L.

Matteuccia struthiopteris (L.) Todaro .... ostrich fern

meadow horsetail .... ***Equisetum pratense*** Ehrh.

Mentha arvensis L. .... wild mint

Menyanthes trifoliata L. .... buck-bean

Mertensia paniculata (Ait.) G. Don. .... tall lungwort

Milium effusum L. .... millet grass

millet grass .... ***Milium effusum*** L.

Mitella nuda L. .... bishop's-cap

mnium moss .... ***Mnium*** spp.

Mnium spp. .... mnium moss

mountain maple .... ***Acer spicatum*** Lamb.

narrow reed grass .... ***Calamagrostis stricta*** (Timm) Koeler

northern bedstraw .... ***Galium boreale*** L.

northern laurel .... ***Kalmia polifolia*** Wang.

northern star-flower .... ***Trientalis borealis*** Raf.

northern willowherb .... ***Epilobium ciliatum*** Raf.

oak fern .... ***Gymnocarpium dryopteris*** (L.) Newm.

one-sided wintergreen .... ***Orthilia secunda*** (L.) House

Orthilia secunda (L.) House .... one-sided wintergreen

Oryzopsis asperifolia Michx. .... rice grass

ostrich fern .... ***Matteuccia struthiopteris*** (L.) Todaro

Oxycoccus microcarpus Turcz. .... small bog cranberry

palmate-leaved coltsfoot .... ***Petasites palmatus*** (Ait.) A. Gray

peat moss .... ***Sphagnum*** spp.

Peltigera aphthosa (L.) Willd. .... studded leather lichen

Petasites palmatus (Ait.) A. Gray .... palmate-leaved coltsfoot

Phalaris arundinacea L. .... reed canary grass

Phalaris sp./***Phragmites*** sp. .... reed grass

Phragmites australis (Cav.) Trin. ex Steud. .... giant reed grass

Picea glauca (Moench) Voss .... white spruce

Picea mariana (Mill.) B.S.P. .... black spruce

pin and choke cherry .... ***Prunus*** spp.

pin cherry .... ***Prunus pensylvanica*** L.f.

Pinus banksiana Lamb. .... jack pine

Plagiomnium cuspidatum (Hedw.) (Kop.) .... woodsy leafy moss

Plagiomnium spp. .... leafy moss

Pleurozium schreberi (Brid.) Mitt. .... Schreber's moss

Poa palustris L. .... fowl bluegrass

Polytrichum commune Hedw. .... common hair-cap

Polytrichum piliferum Hedw. .... awned hair-cap

Polytrichum strictum Brid. .... slender hair-cap

Populus balsamifera L. .... balsam poplar

Populus deltoides Marsh. .... western cottonwood

Populus tremuloides Michx. .... aspen

Potentilla palustris (L.) Scop. .... marsh cinquefoil

prickly rose .... ***Rosa acicularis*** Lindl.

Prunus pensylvanica L.f. .... pin cherry

Prunus spp. .... pin and choke cherry

Prunus virginiana L. .... choke cherry

Ptilium crista-castrensis (Hedw.) De Not. .... knight's plume moss

Pyrola asarifolia Michx. .... common pink wintergreen

ragged moss .... ***Brachythecium*** spp.

red and white baneberry ....***Actaea rubra*** (Ait.) Willd.

red pixie-cup ....***Cladonia borealis*** S. Stenroos

reed canary grass ....***Phalaris arundinacea*** L.

reed grass ....***Phalaris*** sp./***Phragmites*** sp.

reindeer lichen .... ***Cladina mitis*** (Sandst.) Hale & W. Culb.

reindeer lichen .... ***Cladina*** spp.

Rhamnus alnifolia L'Hér. .... alder-leaved buckthorn

Ribes lacustre (Pers.) Poir. .... bristly black currant

Ribes spp. .... currant

Ribes triste Pall. .... wild red currant

rice grass .... *Oryzopsis asperifolia* Michx.

river alder .... *Alnus tenuifolia* Nutt.

Rosa acicularis Lindl. .... prickly rose

rose .... *Rosa acicularis* Lindl., *Rosa woodsii* Lindl.

Rubus chamaemorus L. .... cloudberry

Rubus idaeus L. .... wild red raspberry

Rubus pubescens Raf. .... dewberry

rush .... *Juncus* spp.

Salix bebbiana Sarg. .... beaked willow

Salix spp. .... willow

sand heather .... *Hudsonia tomentosa* Nutt.

saskatoon .... *Amelanchier alnifolia* Nutt.

Schizachne purpurascens (Torr.) Swallen .... false melic

Schreber's moss .... *Pleurozium schreberi* (Brid.) Mitt.

Scirpus spp. .... bulrush

Scirpus validus Vahl .... common great bulrush

Scutellaria galericulata L. .... marsh skullcap

sedge .... *Carex* spp.

Shepherdia canadensis (L.) Nutt. .... Canada buffalo-berry

shield fern .... *Dryopteris carthusiana* (Vill.) H.P. Fuchs

showy aster .... *Aster conspicuus* Lindl.

sickle moss (brown moss) .... *Drepanocladus uncinatus* (Hedw.) Warnst.

slender hair-cap .... *Polytrichum strictum* Brid.

small bog cranberry .... *Oxycoccus microcarpus* Turcz.

small enchanter's nightshade .... *Circaea alpina* L.

Smilacina stellata (L.) Desf. .... star-flowered Solomon's seal

Smilacina trifolia (L.) Desf. .... three-leaved Solomon's seal

snowberry .... *Symphoricarpus albus* (L.) Blake

Solidago canadensis L. .... Canada goldenrod

Solidago gigantea Ait. .... late goldenrod

Sphagnum spp. .... peat moss

spreading dogbane .... *Apocynum androsaemifolium* L.

stair-step moss .... *Hylocomium splendens* (Hedw.) B.S.G.

star-flowered Solomon's seal .... *Smilacina stellata* (L.) Desf.

stiff club-moss .... *Lycopodium annotinum* L.

studded leather lichen .... *Peltigera aphthosa* (L.) Willd.

sweet-scented bedstraw .... *Galium triflorum* Michx.

Symphoricarpus albus (L.) Blake .... snowberry

tall lungwort .... *Mertensia paniculata* (Ait.) G. Don.

tamarack .... *Larix laricina* (Du Roi) K. Koch

Thalictrum venulosum Trel. .... veiny meadow rue

three-leaved Solomon's seal .... *Smilacina trifolia* (L.) Desf.

Tomenthypnum nitens (Hedw.) Loeske .... golden moss

Trientalis borealis Raf. .... northern star-flower

tufted moss .... *Aulacomnium palustre* (Hedw.) Schwaegr.

twin-flower .... *Linnaea borealis* L.

twining honeysuckle .... *Lonicera dioica* L.

Typha latifolia L. .... cattail

Ulmus americana L. .... white elm

Urtica dioica L. .... common nettle

Vaccinium myrtilloides Michx. .... blueberry

Vaccinium vitis-idaea L. .... bog cranberry

veiny meadow rue .... *Thalictrum venulosum* Trel.

Viburnum edule (Michx.) Raf. .... low-bush cranberry

Viburnum opulus L. .... high-bush cranberry

Vicia americana Muhl. .... wild vetch

Viola canadensis L. .... western Canada violet

Viola renifolia A. Gray .... kidney-leaved violet

Viola selkirkii Pursh .... great-spurred violet

western Canada violet .... *Viola canadensis* L.

western cottonwood .... *Populus deltoides* Marsh.

white birch .... *Betula papyrifera* Marsh.

white elm .... *Ulmus americana* L.

white spruce .... *Picea glauca* (Moench) Voss

wild lily-of-the-valley .... *Maianthemum canadense* Desf.

wild mint .... *Mentha arvensis* L.

wild red currant .... *Ribes triste* Pall.

wild red raspberry .... *Rubus idaeus* L.

wild sarsaparilla .... *Aralia nudicaulis* L.

wild strawberry .... *Fragaria virginiana* Duchesne

wild vetch .... *Vicia americana* Muhl.

willow .... *Salix* spp.

wire rush .... *Juncus balticus* Willd.

woodland horsetail .... *Equisetum sylvaticum* L.

woodland strawberry .... *Fragaria vesca* L.

woodsy leafy moss .... *Plagiomnium cuspidatum* (Hedw.) (Kop.)

APPENDIX 8: PLANT IDENTIFICATION REFERENCES

Brodo, I.M.; Hawksworth, D.L. 1877. Alectoria and allied genera in North America. Natl. Mus. Nat. Sci., Natl. Mus. Can., Ottawa, Ontario. Publ. Bot. 6.

Conard, H.S.; Redfearn, P.L. 1979. How to know the mosses and liverworts. Wm. C. Brown, Dubuque, Iowa.

Cormack, R.G.H. 1977. Wild flowers of Alberta. Hurtig, Edmonton, Alberta.

Crum, H.A. 1976. Mosses of the Great Lakes Forest. Rev. ed. Herb., Univ. Mich., Ann Arbor, Michigan.

Crum, H.A.; Anderson, L.E. 1981. Mosses of eastern North America. 2 vols. Columbia Univ. Press, New York, New York.

Egan, R.S. 1987. A fifth checklist of the lichen-forming, lichenicolous, and allied fungi of the continental United States and Canada. Bryologist 90:77-173.

Farrar, J.L. 1995. Trees in Canada. Fitzhenry and Whiteside Ltd., Markham, Ontario and Nat. Resour. Can., Can. For. Serv., Ottawa, Ontario.

Goffinet, B.; Hastings, R.I. 1994. The lichen genus *Peltigera* (Lichenized Ascomycetes) in Alberta. Prov. Mus. Alberta, Edmonton, Alberta. Nat. Hist. Occas. Pap. 21.

Hale, M.E. 1979. How to know the lichens. 2nd ed. Wm. C. Brown, Dubuque, Iowa.

Hitchcock, C.L.; Cronquist, A.; Ownbey, M.; Thompson, J.W. 1955-69. Vascular plants of the Pacific Northwest. 5 vols. Univ. Wash. Press, Seattle, Washington.

Ireland, R.R. 1982. Moss flora of the Maritime provinces. Natl. Mus. Nat. Sci., Natl. Mus. Can., Ottawa, Ontario.

Johnson, D.; Kershaw, L.; MacKinnon, A.; Pojar, J. 1995. Plants of the western boreal forest and aspen parkland. Lone Pine Publ., Edmonton, Alberta.

Lawton, E. 1971. Moss flora of the Pacific Northwest. Hattori Bot. Lab., Japan.

Looman, J.; Best, K.F. 1979. Budd's flora of the Canadian prairie provinces. Agric. Can., Res. Branch, Ottawa, Ontario. Publ. 1662.

MacKinnon, A.; Pojar, J.; Coupe, R. eds. 1992. Plants of northern British Columbia. Lone Pine Publ., Edmonton, Alberta.

Moss, E.H. 1983. Flora of Alberta. 2nd ed. Revised by J.G. Packer. Univ. Toronto Press, Toronto, Ontario.

Schofield, W.B. 1992. Some common mosses of British Columbia. 2nd ed. Royal British Columbia Mus., Victoria, British Columbia. Handb. 28.

Schuster, R.M. 1969. The Hepaticae and Anthocerotae of North America east of the Hundredth Meridian. Vol. 2. Columbia Univ. Press, New York, New York.

Scoggan, H.J. 1978-79. The flora of Canada. 4 vols. Natl. Mus. Nat. Sci., Natl. Mus. Can., Ottawa, Ontario. Publ. Biol. 7.

Stotler, R.; Crandall-Stotler, B. 1977. A checklist of the liverworts and hornworts of North America. Bryologist 80:405–428

Thomson, J.W. 1967. The lichen genus *Cladonia* in North America. Univ. Toronto Press, Toronto, Ontario.

Vitt, D.H.; Marsh, J.E.; Bovey, R.B. 1988. Mosses, lichens, and ferns of northwestern North America. Lone Pine Publ., Edmonton, Alberta.

INDEX OF COMMON AND SCIENTIFIC NAMES OF ILLUSTRATED PLANTS